Sustainable Alternative Syngas Fuel

Edited by Chaouki Ghenai and Abrar Inayat

Published in London, United Kingdom

IntechOpen

Supporting open minds since 2005

Sustainable Alternative Syngas Fuel
http://dx.doi.org/10.5772/intechopen.78190
Edited by Chaouki Ghenai and Abrar Inayat

Contributors
Enrique Posada, Gilmar Saenz, Onoriode Avbenake, Baba Jibril, Faruk Ibrahim Yakasai, Zhang Min, Zhang Haifeng, Mario Toledo, Raghda Ahmed El-Nagar, Alaa Ali Ghanem, Gartzen Lopez, Jon Alvarez, Maria Cortazar, Laura Santamaría, Enara Fernandez, Martin Olazar

First published in London, United Kingdom, 2019 by IntechOpen
IntechOpen is the global imprint of INTECHOPEN LIMITED, registered in England and Wales, registration number: 11086078, 7th floor, 10 Lower Thames Street, London, EC3R 6AF, United Kingdom
Printed in Croatia

British Library Cataloguing-in-Publication Data
A catalogue record for this book is available from the British Library

Additional hard and PDF copies can be obtained from orders@intechopen.com

Sustainable Alternative Syngas Fuel
Edited by Chaouki Ghenai and Abrar Inayat
p. cm.
Print ISBN 978-1-78984-580-8
Online ISBN 978-1-78984-581-5
eBook (PDF) ISBN 978-1-78985-288-2

For EU product safety concerns:
IN TECH d.o.o., Prolaz Marije Krucifikse Kozulić 3, 51000 Rijeka, Croatia, info@intechopen.com or visit our website at intechopen.com.

Meet the editors

Chaouki Ghenai is an Associate Professor at the Sustainable and Renewable Energy Engineering (SREE) Department, College of Engineering; Coordinator of the Sustainable Energy Development Research Group at the Research Institute of Science and Engineering (RISE); and the Chairman of the Research Funding Department at the Office of Vice Chancellor for Research and Graduate Studies, University of Sharjah. Dr. Ghenai has a PhD in Mechanical Engineering from Orleans University, France. Before joining the University of Sharjah, he worked at the Florida Atlantic University, Florida International University, University of Los Angeles California, Kansas State University, and Cornell University. Dr. Ghenai's research interests are renewable energy, energy efficiency, energy conversion systems, combustion, biofuels, alternative fuels, clean combustion technologies, waste to energy, sustainability, eco-design, energy-water nexus, energy planning and climate change mitigation assessment, modeling and simulation of micro gird power systems, and air pollution. Dr. Ghenai has more than 100 papers published in journals, conferences, books, and book chapters.

Dr. Abrar Inayat has been an Assistant Professor in the Department of Sustainable and Renewable Energy Engineering at University of Sharjah since 2015. He received his Ph.D. in Chemical Engineering from the Universiti Teknologi PETRONAS, Malaysia. He completed his Masters in Energy Engineering from Mälardalens University, Sweden. Dr. Abrar has published more than 90 research articles including journal articles and conference proceedings. He has successfully completed many research projects by university-industry collaboration and has obtained several international awards. His main research fields are renewable and sustainable energy systems, biomass and bioenergy, process system engineering, production of biofuels, modelling, simulation and optimization of biomass processing technologies, process integration, CO2 capture and utilization, heat integration, energy management, and hybrid energy systems.

Contents

Preface

The combustion of conventional fossil fuels is the largest contributing factor to the greenhouse gases in the atmosphere. The development and use of sustainable and alternative fuels (syngas, biogas, biodiesel, bio-oil, hydrogen) derived from sources other than petroleum is needed due to the limited fossil fuel resources, the need for reduction of atmospheric greenhouse gas emissions, energy security, and to meet the future high energy demand due to population growth. New alternative fuels that can be produced locally and derived from renewable sources will be more sustainable compared to fossil fuels. Alternative and renewable fuels can be produced using different thermochemical and bio-chemical processes. Gasification is a thermochemical process used to produce syngas fuel (mainly hydrogen and carbon dioxide) from renewable (biomass) and conventional (coal) sources. The syngas fuels produced from the gasification process can be used for different applications: power generation (combustion of syngas fuel in gas turbine engines), heating, and transportation (internal combustion engines).

This book intends to provide the reader with a comprehensive overview of the current technologies, methods, and strategies of syngas fuel production, characterization, and application.

Chaouki Ghenai
Associate Professor,
Sustainable and Renewable Energy Engineering Department (SREE),
College of Engineering,
Coordinator - Sustainable Energy Development Research Group,
Research Institute of Science and Engineering (RISE),
Chairman – Research Funding Department, VCRGS
University of Sharjah,
Sharjah, United Arab Emirates

Abrar Inayat
Assistant Professor,
Sustainable and Renewable Energy Engineering Department (SREE),
College of Engineering,
Member of the Sustainable Energy Development Research Group,
Research Institute of Science and Engineering (RISE),
University of Sharjah,
Sharjah, United Arab Emirates

Section 1

Syngas Fuel

Chapter 1

Waste to Energy and Syngas

Enrique Posada and Gilmar Saenz

Abstract

Getting energy from waste is one of the best alternatives for sustainable handling of waste. Mass burning is generally the preferred option. Usually, this applies to large facilities where more than 500 tons of waste per day are treated. Syngas production from waste has also been tried with mixed success. This chapter reviews the situation in this field and proposes an alternative based on co-combustion with coal as a possible route, applied preferably to treat municipal solid waste (MSW) and biosolids from small- or medium-sized municipalities, producing less than 200 tons of waste per day, with the aim of generating electric energy. For this, a theoretical model is proposed and applied to a specific case.

Keywords: waste to energy, municipal solid waste, design, modeling, syngas composition, technologies, experience, electric energy, coal, co-combustion

1. Introduction

This chapter deals with the possibilities of making use of municipal solid waste (MSW) in combined gasification systems with coal to help solving two situations. One is the need for a more sustainable use of highly available coal resources and the other is the need for a more sustainable handling of domestic solid wastes, which are not properly disposed. When these two combine, as is the case for a country like Colombia, there are real spaces for the use of waste to energy technologies.

Coal is an abundant resource in many places of the world. Unfortunately, the combustion of coal has been clearly associated with the generation of CO_2 and global warming, which has caused a tendency to gradually abandon coal as an energy resource, preferring natural gas and renewable energy. This is a worrying situation for a country like Colombia, which possess very large coal deposits. Currently, this country is exporting large amounts of coal and this contributes largely to the generation of income. In this sense, it is important to find applications for coal, both in chemical process and more sustainable energy systems and also develop ways for CO_2 recovery and conversion that allow for the continuous use of coal.

The waste problem is very important in developing countries like Colombia [1]. With 49 million people in 2017 and its population mostly concentrated in the Andean highlands and along the Caribbean coast, it has 31 cities of more than 200,000 habitants and 65 with more than 100,000; being one of most urbanized countries in the region, its urban population is estimated at 76%. Informality and poverty are big problems, and these come associated with informal waste recycling practices. With a medium generation of 0.54 kg/hab./day, the estimated daily generation is around 26,000 tons. Colombia is a model in the region in the recycling of paper and cardboard, with a recovery of 57%. This has to do with the existence of industrial plants able to use these materials in their process, which has favored a well-organized recycling scheme. Currently in the country, the recycling rate of waste such as paper, cardboard, glass, metals, and plastics is 17%, and by 2019, the goal will be to achieve a recycling target of 20% as a result of the implementation of regulatory instruments in the public cleaning services and the tariff frameworks, processes that the national government advances. The rest of the waste goes to waste dumps or sanitary landfills as there are not any thermal treatment facilities in the country. Very few of the landfills facilities have water lixiviate treating plants or methane burning systems. Space is becoming an issue and there are growing concerns and limitations about the growth of the landfill system areas in the coming years. In other cases, environmental concerns are becoming more and more important [2–5].

Waste to energy systems are very important for the sustainable disposition of municipal waste as has been consistently shown in developed countries. This has to do with available technology. In general, in developing countries, there is lack of companies that can manufacture equipment for thermal treatment systems capable of handling hundred or thousand tons per day of mixed waste, burning them in a controlled way, generating electricity, and controlling the air pollution problems related to this. This means that local responsible waste-handling entities will tend to look for solutions with external providers and this means usually very high initial investments. As shown in the case of China and India, this can be changed, creating competitive sectors in the WtE technology, able to confront their own situations and to export technology and equipment.

Engineering and design are very important components of the necessary technology for the development of WtE (waste to energy) systems in a country. Implementing these systems requires detailed studies and planning activities and it is advisable to do the projects considering all the engineering stages. There is always the temptation and the idea that the projects can be accelerated and put into place based on the experience and support of suppliers and makers, by means of EPC developments. The idea being that in such a way, the engineering stages can be simplified or even avoided. This normally is a much costlier and rigid solution and does not contribute to developing local technology and desired prosperity. In the solution of the problems, there is ample space to develop a region, as compared to relying only on externally provided solutions.

One of the most important stages is the development of conceptual studies and engineering based as much as possible on local expertise, duly backed, of course with external experience and support. The authors are part of an international working group known as WTERT supported by Earth Institute at Columbia University [6]. The Waste to Energy Research and Technology Council (WTERT) brings together engineers, scientists, and managers from universities and industries worldwide and the authors belong to the Colombian chapter, which is supported by ACIEM (Engineering Colombian Association). WTERT tries to identify and advance the best available waste to energy technologies for the recovery of energy or fuels from municipal solid wastes and other industrial, agricultural, and forestry residues. The authors are also project engineers at HATCH, an international engineering company, and have experience in waste to energy systems for industrial applications.

As part of their work, they participated in a project aimed at using gasification systems based on the co-combustion of coal with biosolids coming from a municipal water treatment system [7–9]. This chapter considers using this technology for waste to energy systems applied to municipal solid waste (MSW). It reviews the situation in this field. This, in order to explore the basis for an alternative based on co-combustion with coal for generating syngas in small- or medium-sized municipalities, produces less than 200 tons of waste per day. It develops a theoretical model applied to the specific case of municipal waste similar to the one generated at the city of Medellin, where the authors work, co-gasified with available local coal.

Gasification processes involve the reaction of carbonaceous feedstock with an oxygen-containing reagent, usually oxygen, air, steam, or carbon dioxide, generally at temperatures in excess of 800°C. It involves the partial oxidation of a substance which implies that oxygen is added but the amounts are not sufficient to allow the fuel to be completely oxidized and full combustion to occur [10]. The main product is syngas, which is a mixture of gases including CO and H_2, which can be used to produce fuels and chemicals, or be burned to generate heat or electricity. Some by-products are ash and tars depending on the technology used.

The basics of the gasification process can be found in many publications and books. MSW gasification has been an object of many studies also and the process details and specificities have been compiled and documented. Zafar [10] shows the qualitative basics, advantages, and disadvantages, as well as classifications depending on the technology, feedstock, and reactors, focused on municipal solid waste. Arena [11] presents a deeper treatment of the gasification technology, the chemistry, reactor and technology description and comparison, and environmental aspects. In his thesis, Klein [12] also analyzes these aspects in depth and also considers investment and operative costs with data of operating plants at that time.

In terms of co-gasification, specific studies have been carried out showing the technical feasibility of the technique, and quantifying the improvements depending on the co-gasification agent.

Koukouzas et al. analyzed co-gasification of MSW with coal. They evaluated the techno-economic feasibility, of a 30-MW (e) co-gasification power plant based on integrated gasification combined cycle (IGCC) technology, using lignite and refuse-derived fuel (RDF), in the region of Western Macedonia, Greece. The preliminary cost estimation indicated that this plant was not profitable, due to high specific capital investment and in spite of the lower fuel supply cost. The estimated cost of electricity was not competitive, compared to the dominating prices for the Greek electricity market [13].

Hu et al. studied a three-stage system for co-gasification of MSW with high-alkali coal char. Tar content was controlled to as low as 11.3 mg/Nm^3 and HCl to 17.6 mg/Nm^3. Lower heating value attains 12.2 MJ/Nm^3, meeting the intake-gas conditions for internal combustion engines. They concluded that high-quality syngas can be produced at a steady yield rate of 1.57 Nm^3/kg from three-stage gasifier, due to dichlorination and catalytic tar cracking action of high-alkali coal char at a low cost [14].

Co-gasification of MSW with switchgrass cuttings, by means of a small commercial-scale downdraft gasifier (100 kg/h), indicates that co-gasification of up to 40% MSW performed satisfactorily. The heating values of syngas were 6.2, 6.5, and 6.7 MJ/Nm^3 for co-gasification ratios of 0, 20, and 40%, respectively; in the same cases, the cold and hot gas efficiencies were 60.1, 51.1, and 60.0% and 65.0, 55.2, and 64.4% [15]. Eghtedaei et al. also analyzed co-gasification with biomass and found an improvement in the H_2 concentration [16].

The co-gasification with the bottom ash has been studied, finding improvements in the final ash quality and the gas emissions without important changes in the operability and syngas quality [17].

These few examples show that in principle, not only MSW gasification, but also co-gasification are feasible at different scales, including commercial scale. Many companies or institutes have developed their own process routes with particularities to be more efficient or suitable for the feedstock. In addition to the studies reviewed, some other successful cases could be considered.

Enerkem has effectively developed its own process to obtain methanol and ethanol from MSW through gasification and has an operating plant in Alberta,

Canada [18]. Mitsubishi Heavy Industries has a medium-sized plant in Kushiro, Japan, which has been operating since 2006, processing 240 T/day of MSW (2 units × 120 T/day), producing 4.6 MW of electricity. Their technology includes an ash melting system that improves the ash quality and controls the dioxin emissions [19]. Currently, Fulcrum-Bioenergy is preparing the construction of a MSW gasification facility in Nevada (USA) to produce 10 million gallons a year of biofuels [20]. Aries Clean Energy has different facilities already working in the USA. In Sanford, Florida, they installed a fluidized bed gasification plant for 30 T/day biosolids from a sewage treatment plant [21]. In Lebanon, Tennessee, a downdraft reactor gasifies 64 T/day of biomass to produce heat that is used with organic Rankine cycles (ORCs) [22]. The same technology was used in Covington, Tennessee, with a reactor of 12 T/day mixture of wood residues and sludge moving a 235-kW ORC [23]. In Boral Bricks, Alabama, 12 modular downdraft systems were used to process residual wood to produce syngas to be burned in brick furnaces [24].

Tanigaki et al. have reviewed the operation of two plants in Japan. They reported more than 46 gasification facilities working nowadays in Japan but focused on the two more recent ones, one processes MSW with higher operating hours and lower consumables in Japan. The other one is focused on its waste flexibility, processing not only MSW but also IBA, rejects from recycling center, and sewage sludge. They show the reliability of these plants as well as their effectiveness on the MSW treatment, energy efficiency, and accomplishing environmental requirements [25].

There are many gasification facilities in the world. A good review of them can be found in the Worldwide Syngas database of the Global Syngas Technology Council [26]; here, the facilities can be located and filtered by feedstock, product, and technology among others. In the following studies, in addition to very good technological reviews of the MSW thermal treatment, especially on gasification, there are sets and lists of plants, facilities around the world with their capacities and owners.

- Thermal municipal solid waste gasification [27].
- Thermal processing of waste [28].
- Municipal solid waste (MSW) to liquid fuels synthesis, volume 1: Availability of feedstock and technology [29].
- Feasibility study on solid waste to energy: Technological aspects [30].
- Gasification of non-recycled plastics from municipal solid waste in the United States: Thermal municipal solid waste gasification [31].
- Thermal plasma gasification of municipal solid waste (MSW) [32].

There can be found good examples of feasible and working projects for MSW treatment; however, it is important to note that these projects have specific and contextual difficulties. Hakan Rylander, an experienced actor in WtE, is a bit skeptical about gasification of MSW, mostly because of the heterogeneity of the feedstock, and because the energy balance sometimes has turned out to be negative [33]. Also, Tangri and Wilson [34], make an interesting risk analysis of the gasification and pyrolysis of MSW. They conclude that "the potential returns on waste gasification are smaller and more uncertain, and the risks much higher, than proponents claim," "Technical and economic challenges for gasification projects include failing to meet projected energy generation, revenue generation, and emission targets. Gasification plants also have historically sought public subsidies to be profitable." At the end of the document, there is a list of ten notable cases of plants and facilities around the world that have stopped operations.

There is no general rule to assure success of a MSW gasification or co-gasification facility; it depends on the technology used, the nature and variability of the feedstock, and strongly on the local cost and price structure. Where landfilling is still cheap and permitted, WtE tends to be not an economically feasible option. But where waste disposal is becoming more regulated and costly, a WtE plant of this kind is a great option to reduce the amount of material disposed and its inertness while having a benefit, that could be the obtention of energy or of value-added chemicals.

2. Modeling of municipal solid waste and coal co-combustion to generate syngas

This section develops a theoretical model applied to the specific case of municipal waste. The basic information for this is the composition of the MSW and of the coal to be used, plus their heat powers. **Tables 1** and **2** show the data used. These tables have been prepared by authors based on several studies made during their work with coal boilers and power plants at Colombia. Two cases are considered for the waste. In the first one, waste as currently generated, the average quality of the MSW is considered in the city of Medellin, which is quite rich in organic materials and, so, very high in water content. In the second case, previously separated waste is

Water content	% wet basis	7.20
Carbon	% dry basis	68.77
Hydrogen	% dry basis	4.55
Nitrogen	% dry basis	1.27
Oxygen	% dry basis	12.08
Sulfur	% dry basis	0.45
Ashes	% dry basis	12.87
High heat value (dry basis)	KJ/kg	25,911
Lower heat value (wet basis)	KJ/kg	23,155

Table 1.
Coal properties considered [35].

Case		As generated	Separated
Water content	% wet basis	45.58	24.93
Carbon	% dry basis	42.70	38.50
Hydrogen	% dry basis	5.93	5.35
Oxygen	% dry basis	37.95	34.22
Ashes	% dry basis	13.42	21.93
High heat value (dry basis)	KJ/kg	16,244	14,647
Lower heat value (wet basis)	KJ/kg	8,129	10,111

Table 2.
MSW properties considered [35].

considered, removing 75% of organic material, 50% of paper, 20% of plastics, 55% of glass, 60% of cardboard, and 50% of metals of the generated waste. This would amount to 45% of the initial as generated MSW.

Gasification is modeled considering three combinations for co-gasification, identified by the mass ratio of coal to MSW: 0, 0.25, and 0.50. Saturated steam was supplied at 4 bar relative pressure (ambient pressure 1 bar) with steam-to-MSW mass ratios between 0.0 and 1.0 and heated air (120°C) was supplied with air-to-MSW rates between 1.70 and 5.0. **Figure 1** schematizes the basic model used.

The following chemical reactions were considered for the equilibrium calculations in the simulations. No methane generation was considered. Sulfur was controlled by the addition of calcium carbonate at a mass ratio of 0.0163 to coal.

$$C + CO_2 \leftrightarrow 2CO \quad (1)$$

$$CO + H_2O \leftrightarrow CO_2 + H_2 \quad (2)$$

$$H_2 + 1/2O_2 \leftrightarrow H_2O \quad (3)$$

$$C + H_2O \leftrightarrow CO + H_2 \quad (4)$$

$$C + 1/2O_2 \leftrightarrow CO \quad (5)$$

$$CO + 1/2O_2 \leftrightarrow CO_2 \quad (6)$$

$$C + O_2 \leftrightarrow CO_2 \quad (7)$$

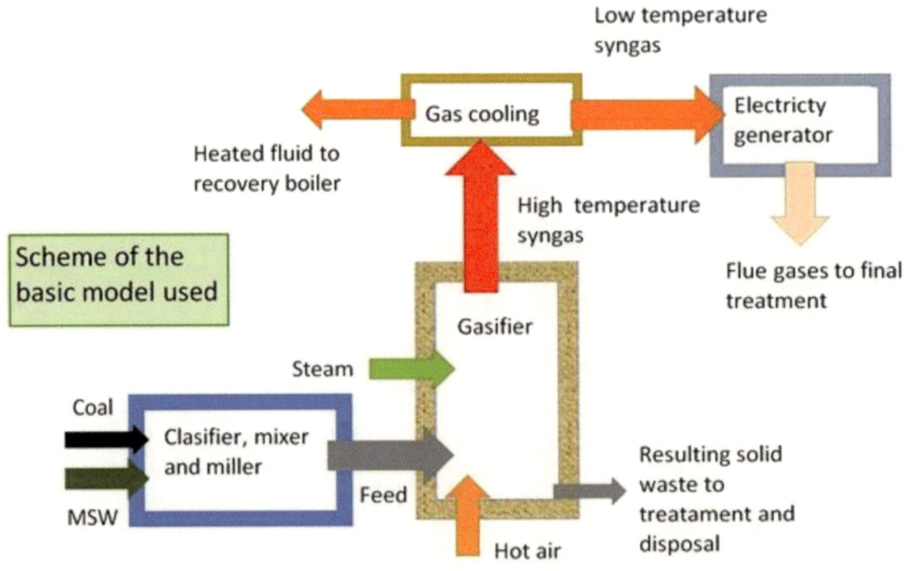

Figure 1.
Scheme of the basic model used.

An iterative model calculation was developed using the solver routine of MS excel in which the concentrations of syngas were iterated with temperature until the expected convergence was found with species mass balance, energy balance, and chemical equilibrium.

Iterations were performed as follows:

- Final syngas temperature is assumed.
- Volumetric fractions of CO_2, CO, H_2, and H_2O in syngas are assumed.
- Fraction of C converted as per reactions (1), (4), and (5) are assumed.
- Fraction of O_2 converted as per reaction (3) and forming CO are assumed.
- Fraction of CO converted as per reaction (2) is assumed.
- With the partial fractions of syngas, equilibria constants for reactions (1) to (7) are found.
- With syngas temperatures, equilibria constants for reactions (1) to (7) are also found.
- A convergence limit was established for the comparison of these two equilibria constants. This was set as less than 15% maximum error for each reaction.
- Mass balance was checked for each species with a convergence limit of less than 5%.
- Energy balance was performed comparing energy formation based on reactions (1) to(7), outgoing syngas enthalpy, incoming vapor and air enthalpy and heat losses (sensible heat, wall and ashes loses). A convergence limit of 5% was established.

Energy formations (kJ/kmol) used were as follows for syngas forming reactions.

$$C + 2H_2 \leftrightarrow CH_4(g), -74.520$$

$$H_2 + 1/2O_2 \leftrightarrow H_2O(g), -241.818$$

$$C + 1/2O_2 \leftrightarrow CO(g), -110.525$$

$$C + O_2 \leftrightarrow CO_2(g), -393.509$$

Enthalpy of syngas was calculated based on syngas composition and specific heat values for each component, depending on temperature, using the expressions of the form: Cp/R = A + B·T + C·T2 + D·T − 2; T (K) where A–D are constants for each gas component and R is the universal gas constant.

Figures 2–12 show the results of the iterations for all major resulting variables. Comments are included for them.

Syngas temperatures tend to increase with higher coal-to-MSW ratios. For each ratio, there is a characteristic curve which indicates higher temperatures for lower air-to-MSW ratios and lower temperatures for higher steam-to-MSW ratios. Temperatures tend to be higher for the case of the separated MSW. **Figure 1** indicates the real working ranges for the simulations. With no coal use, the only range of air-to-MSW ratios that gave convergence in the simulations was in the neighborhood of

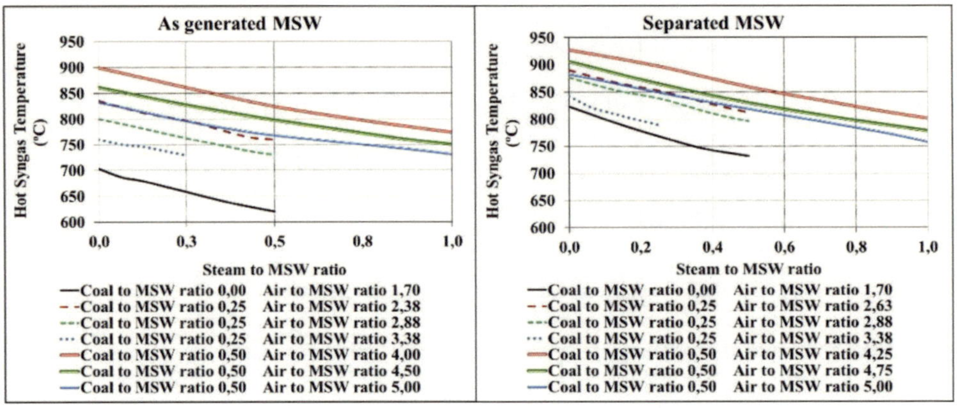

Figure 2.
Resulting syngas temperature.

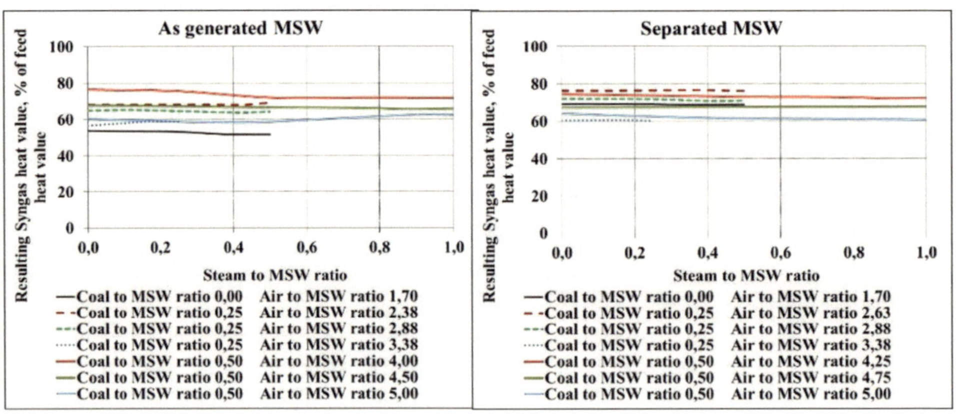

Figure 3.
Resulting heat value in syngas as % of feed heat value.

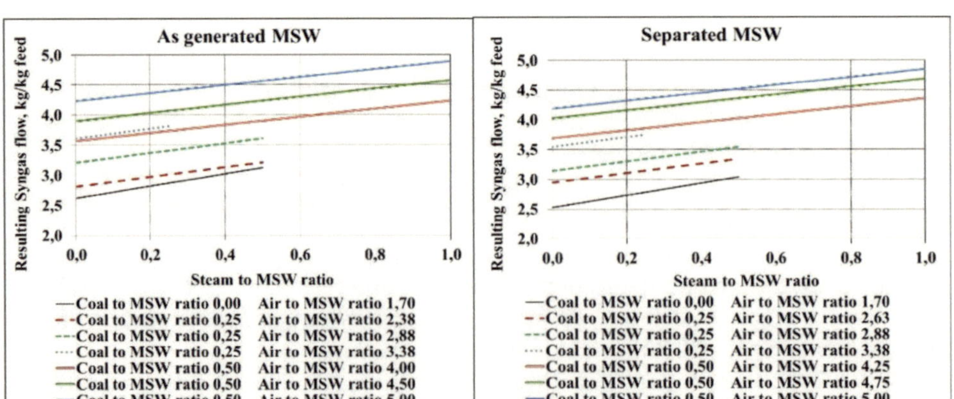

Figure 4.
Syngas flow, kg/kg feed.

1.70. At higher coal-to-MSW ratios, the air-to-MSW ratio can be higher, all the way to 5.0. Syngas temperatures will be between 600 and 940°C.

Syngas heat values tend to increase for higher coal-to-MSW ratios, but this was not entirely consistent. Syngas heat value simulations showed percentages between 60 and 80% of feed heat value and this does not change with steam-to-MSW ratios and tends to decrease with air to MSW ratios.

Syngas flow is linearly related to the studied variables. It increases with air-to-MSW ratio and with steam-to-MSW ratios. The values for the simulated range oscillate between 2.5 and 5.0 kg of syngas per kg of feed. The syngas flow is, basically, the result of adding the incoming flows, discounting the ash emissions. The behavior and the ranges are quite similar for both situations of MSW studied.

As shown in **Figure 5**, syngas heat value is quite independent of steam-to-MSW ratio. It increases with air-to-MSW ratios and, of course, with coal-to-MSW ratios. As compared to the MSW's lower heat value, it tends to be lower, as expected, for the case of no coal co-gasification. Maximum values tend to be double as compared to MSW heat value, obviously because of the impact of coal co-gasification. The values in **Figure 5** are consistent with the ones shown in **Figure 3**. **Figure 6** shows the total energy content of the syngas, adding its heat value to the sensible heat associated to syngas temperature. Those two amount to a value close to the energy value coming from the total feed. It must be said that the incoming hot air and the

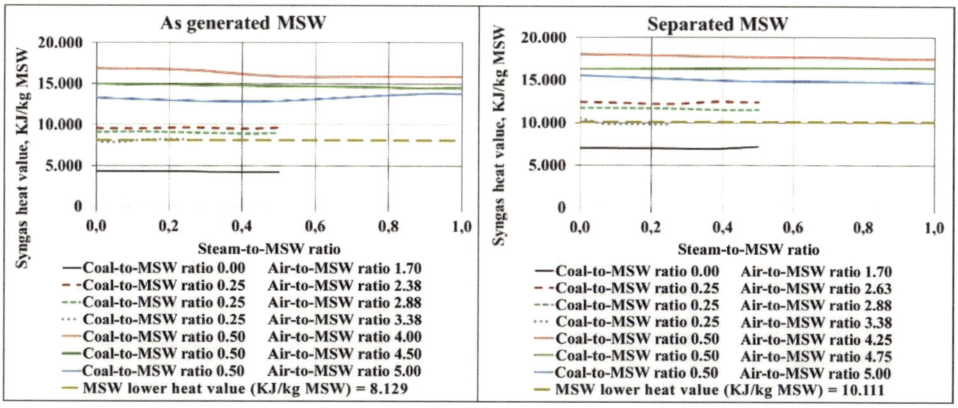

Figure 5.
Syngas heat value, kg/kg MSW.

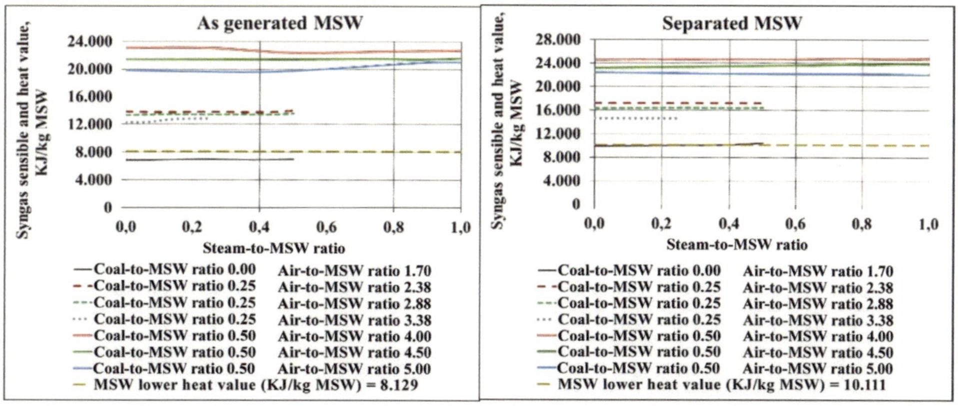

Figure 6.
Syngas heat value and sensible heat, kg/kg MSW.

steam contribute with some energy also, which adds to the outgoing syngas heat value and sensible heat.

The behavior of the total energy in the syngas (**Figure 6**) is quite similar to the behavior of the heat value of **Figure 5**. The heat value corresponds to the chemical (combustion potential) energy associated to H_2 and CO in the syngas.

Some calculations were carried out in the model to determine the potential of syngas to generate electricity. First, the sensible heat potential was determined based on the hot temperature of the syngas. This can be used to generate mechanical work and electricity removing the sensible heat (lowering the temperature, as indicated in **Figure 1**) in a cycle similar to a Rankine cycle. To determine the potential for this, a Carnot cycle's efficiency was calculated using as hot temperature the syngas temperature and as cold temperature the ambient value (25°C). With this Carnot efficiency, an estimation was obtained of a real efficiency based on existing Rankine cycles in which it is possible to get about 35% of the Carnot efficiency. The second estimation was based on expecting an efficiency of 30% for the cycle that employs the combustion heat value of the syngas. This, considering that it could be taken to an internal combustion engine. Combining these two efficiencies, in proportion to the existing contributions (that of heat value and that of sensible heat in the energy content of the syngas), it was possible to estimate the total efficiency

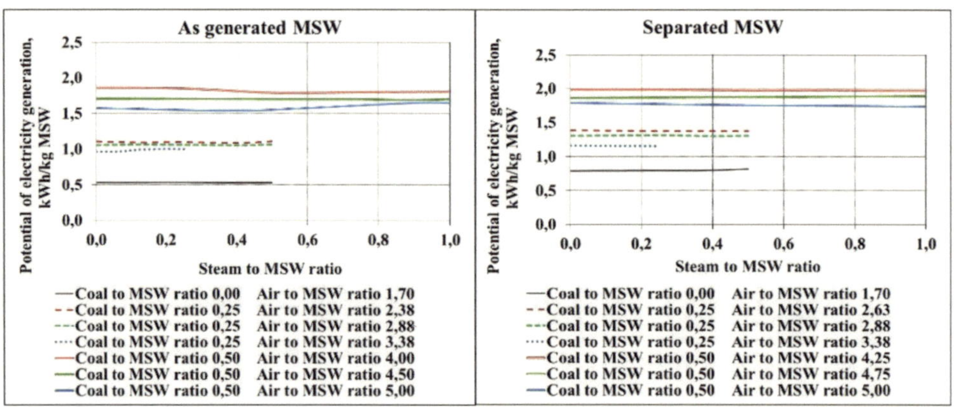

Figure 7.
Potential for electricity generation, kW/kg MSW.

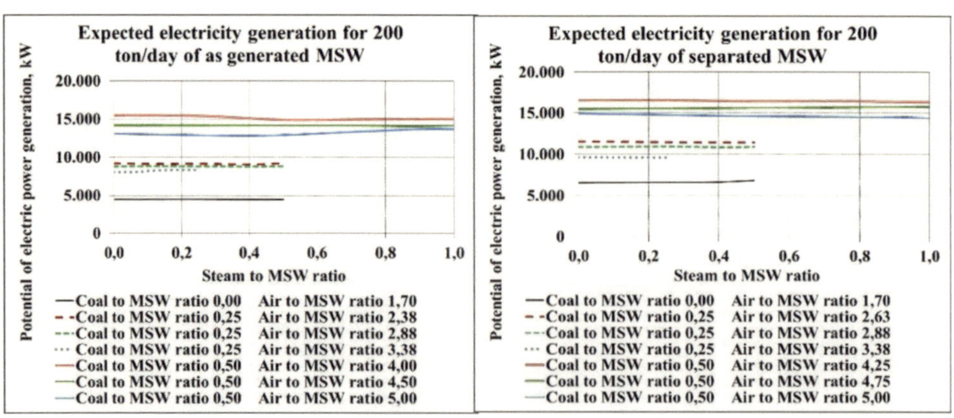

Figure 8.
Electricity generation, in kW, for the processing of 200 tons per day of MSW.

of transformation to electricity and the total potential for electricity generation, which appears in **Figure 7**.

This potential is not affected by steam-to-MSW ratios. It is highly dependent, of course, on coal-to-MSW ratio and it is higher for lower air-to-MSW ratios. The potentials are higher for the case of separated MSW (between 0.75 and 2.2 kW per kg of MSW as compared to a range between 0.5 and 2.0 kW per kg of MSW for the as generated MSW case).

With these potentials, it is possible to estimate the expected electrical generation for a given flow of MSW. **Figure 8** shows the results for a plant processing 200 tons of MSW per day.

These capacities will be between 4800 and 16,000 kW for the as generated MSW and between 6500 and 17,000 kW for the separated MSW. They are not affected by steam-to-MSW ratio, increase clearly with coal-to-MSW ratio, and decrease with air-to-MSW ratio. The ranges indicated in the graphs correspond to the ones for which convergence was found in the iterations, as already mentioned. These plants could generate amounts of electricity quite useful for a given small city in a country like Colombia. Considering a generation of solid waste (as generated) of 0.50 kg/day per habitant, the plant would produce the amounts indicated in **Table 3** for the cases considered. The table compares these figures to the electric consumption of a country like Colombia, estimated at 3.90 kWh per day per capita.

Finally, the simulations permitted to obtain the expected composition of the syngas which will be shown in the next figures.

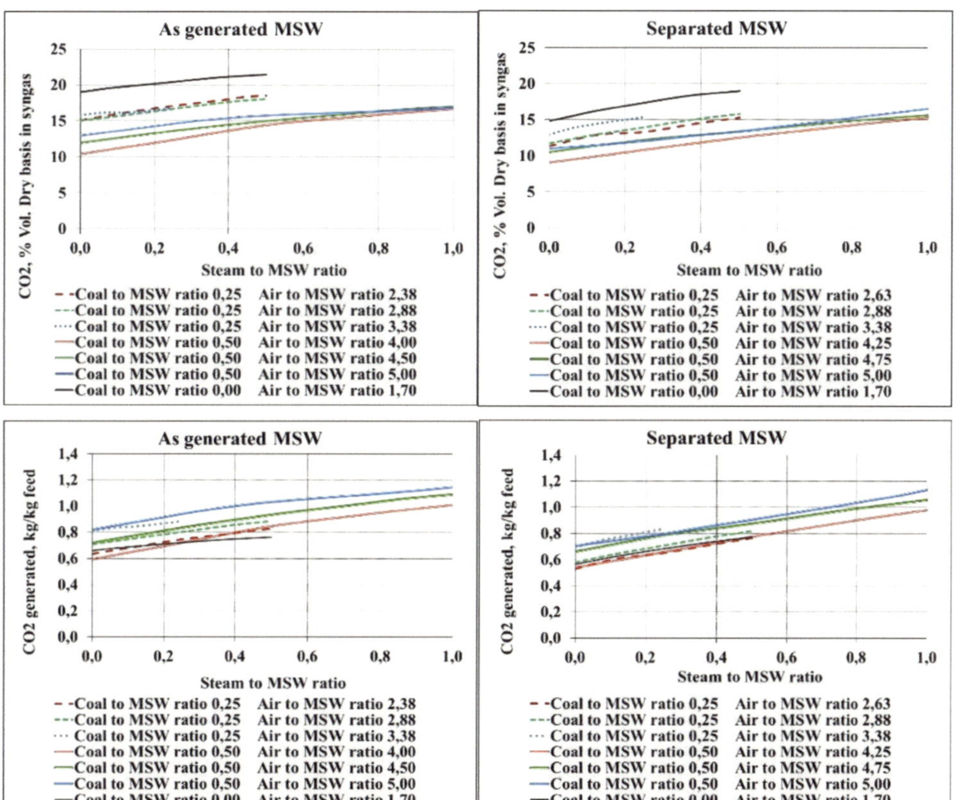

Figure 9.
Syngas CO_2 concentrations and specific emissions.

CO_2 specific emissions increase with steam-to-MSW ratios, with air-to-MSW ratios, and with coal-to-MSW ratios, although in this case depending on the air-to-MSW ratios. Specific emissions are quite similar for both MSW cases.

CO_2 concentrations show a similar behavior but their concentrations in the syngas tend to be somewhat lower for the case of the separated MSW.

CO specific generations decrease with steam-to-MSW ratios and also with air-to-MSW ratios and increase with coal-to-MSW ratios. Specific generations are

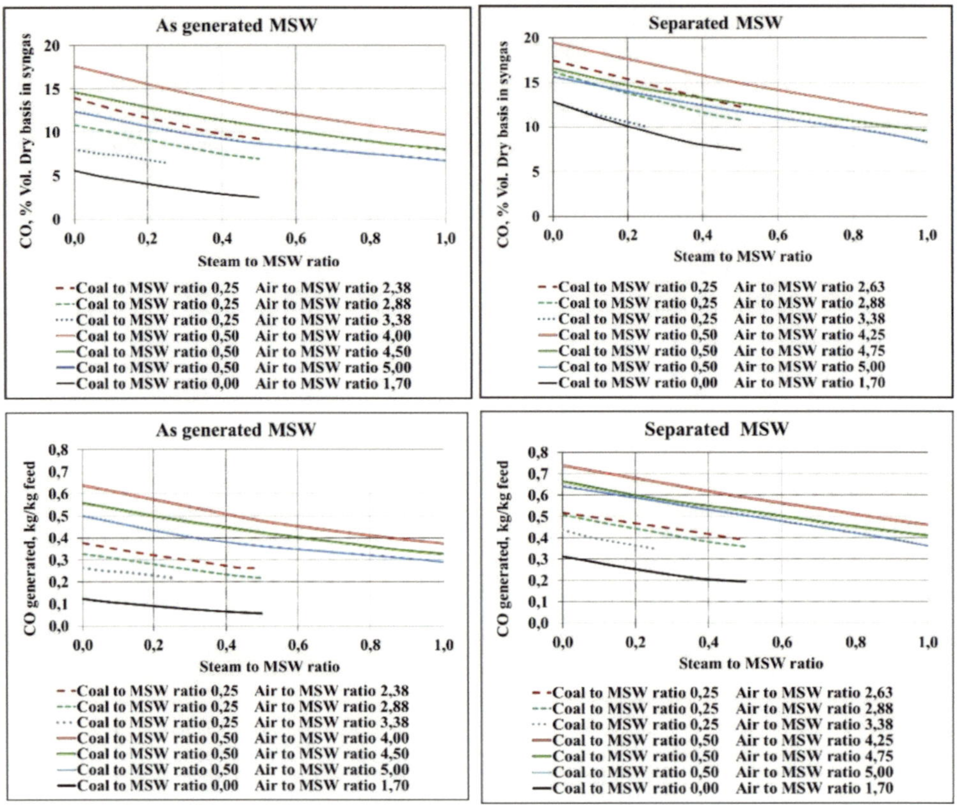

Figure 10.
Syngas CO concentrations and specific generations.

Parameter	Units	As generated	Separated
MSW in Colombia	kg/person day	0.50	0.24
Electricity generated—low	kWh/kg MSW	0.55	0.70
Electricity generated—high	kWh/kg MSW	1.80	2.00
Electricity generated—low	kWh/kg person-day	0.28	0.17
Electricity generated—high	kWh/kg person-day	0.90	0.49
Average electricity consumption in Colombia	kWh/kg person-day	3.90	
Electricity generated—low	% of national use	7.05	4.38
Electricity generated—high	% of national use	23.08	12.51

Table 3.
Per capita electricity generation potential with syngas plants for the considered cases in Colombia.

higher for the case of the separated MSW. CO is one of the two important components of syngas and contributes to its heat value.

CO concentrations show a similar behavior and their concentrations in the syngas tend to be somewhat higher for the case of the separated MSW.

Figure 11 shows the behavior for the H_2 gas as a component of syngas, also one of its two important components and a major contributor to its heat value.

H_2 specific generations increase with steam-to-MSW ratios. This indicates the impact of the conversion of steam to H_2. They decrease also with air-to-MSW ratios. The impact of coal-to-MSW ratios is not entirely clear and is different for the two MSW cases considered. Specific generations are higher for the case of the separated MSW, especially for the case in which no coal is used.

H_2 concentrations show similar behavior and their concentrations in the syngas tend to be somewhat higher for the case of the separated MSW. Concentrations tend to be higher for the low coal-to-MSW ratios.

The water content in the syngas generated with MSW tends to be high, due to the high humidity of the MSW, as shown in **Figure 12**.

H_2O specific generations increase with steam-to-MSW ratios. This indicates a direct relationship coming from the steam added, which is to be expected. They decrease also with air-to-MSW ratios. The impact of coal-to-MSW ratios is evident. When adding coal, the water generation diminishes, as the coal water content is much lower than the one in MSW. Specific generations are clearly lower for the case of the separated MSW, again something to be expected given the lower water content for separated MSW.

H_2O concentrations show a similar behavior in relationship of the direct impact of the steam-to-MSW ratios. The influence of the air-to-MSW ratio is very small.

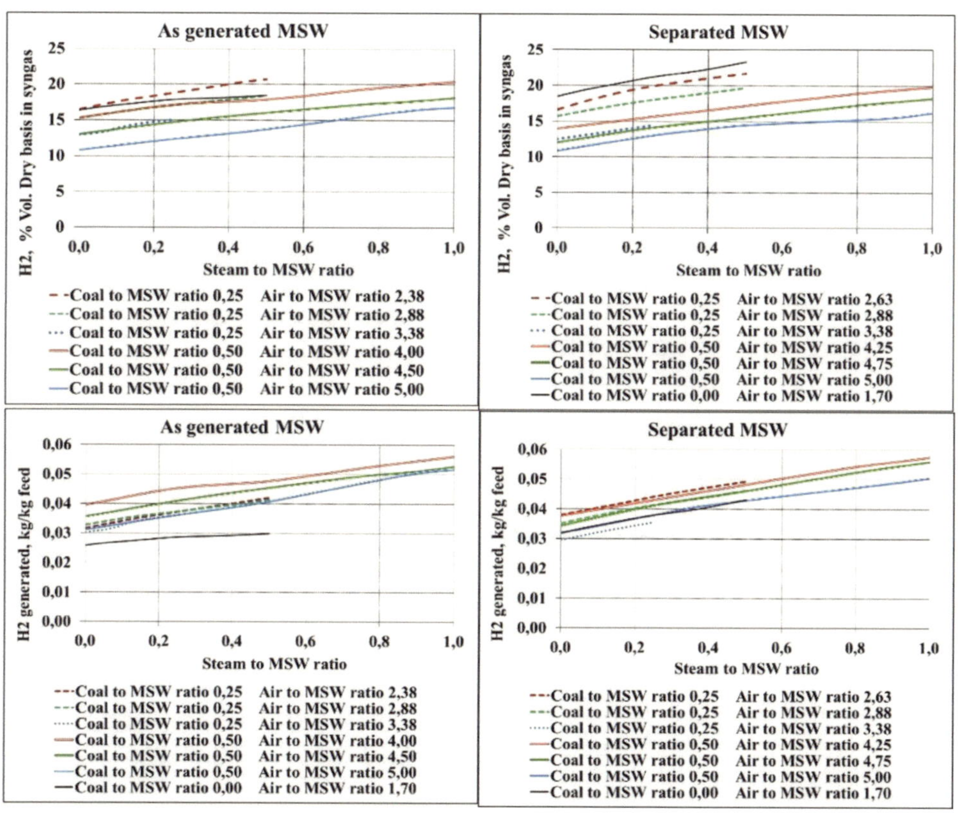

Figure 11.
Syngas H_2 concentrations and specific generations.

The impact of coal-to-MSW ratios is evident as already said. When adding coal, the water generation diminishes as the coal water content is much lower than the one in MSW and their concentrations in the syngas tend to be clearly lower for the case of the separated MSW for the same reasons.

The water content of the syngas has an impact that should be considered in the options for its use. The water concentrations are so high that there could be possibilities of having water condensations on the gases if they reach the dew point

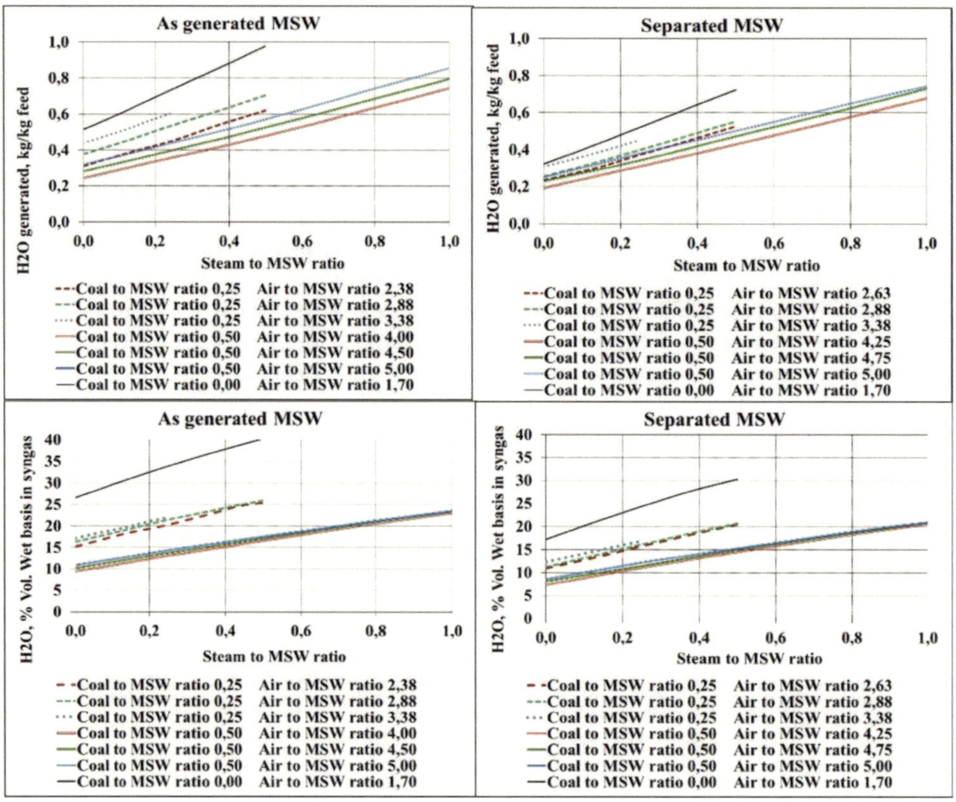

Figure 12.
Syngas H_2O concentrations and specific generations.

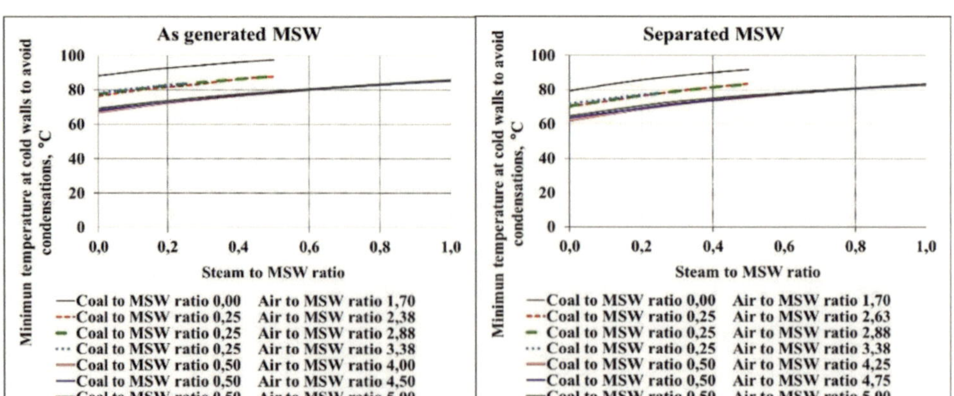

Figure 13.
Minimum cool wall temperatures to avoid water condensation.

temperatures, which could occur at low process temperatures near cold areas, for example in the walls of cooling or transportation equipment. To study this, simulations were made of the wet bulb temperatures assuming cooling under constant total pressure and getting the corresponding saturation temperatures. This simulation is presented in **Figure 13**.

The minimum cool wall temperatures estimated in **Figure 13** include a protection of 20°C, over the calculated dew point temperatures. The dew point temperatures were estimated using psychrometry. The minimum temperatures increase with steam-to-MSW ratio, and decrease with coal-to-MSW ratios, as should be expected. The air-to-MSW ratio did not influence significantly. Temperatures are lower for the case of the separated MSW as expected.

These minimum temperatures can be guaranteed with adequate insulation of the processing equipment and pipe walls for the systems handling the syngas.

3. Conclusions

The theoretical model showed quite consistent results. It was possible to develop a way of estimating syngas characteristics for the gasification of MSW in co-gasification, within practical working ranges for the studied variables. This was done under two extreme conditions for the MSW: as generated in a town with high organic material content and after separation of 55% of the initial waste for recycling and organics treatment (e.g., by biological composting and digestion). The model allowed to find the working ranges for steam-to-MSW ratios (between 0 and 1.0); air-to-MSW (between 1.7 and 5), for co-gasification with coal; and cola-to-MSW ratios in the range of 0.0–0.5.

The gasification can generate electricity in all these ranges, with potentials that go from 0.5 to 2.2 kWh per kg of MSW. For the case of a plant processing 200 tons of MSW per day, the generation capacities would be between 4800 and 17,000 kW. These capacities are entirely within the electricity needs of a country like Colombia. They are between 0.28 and 0.90 kWh per person per day, for the current per capita MSW generated in the country. These figures are to be compared to the current daily electricity per capita use, which is 3.90.

From the practical point of view, it is important to use this as a conceptual basis for future work seeking indications on systems that could be feasible. This will help doing the correct steps. Engineering and design are very important components of the technology necessary to impulse WtE in a country. These systems require detailed studies and planning activities and it is advisable to do the projects considering all the engineering stages. There is always the temptation and the idea that the projects can be accelerated and put into place based on the experience and support of suppliers and makers. This by means of EPC developments, in such a way that engineering stages can be simplified or even avoided. This normally is a much costlier and rigid solution and does not contribute to developing local technology and prosperity. With regard to the solution of the problems, there is ample space to develop a region, as compared to relying only on externally provided solutions. MSW co-gasification with coal seems to be a possible alternative.

Acknowledgements

The authors express their gratitude to HATCH, WTERT—Colombia, ACIEM, and to the Earth Institute at Columbia University and its Waste to Energy Research and Technology Council (WTERT).

Author details

Enrique Posada* and Gilmar Saenz
HATCH, Medellin, Colombia

*Address all correspondence to: enrique.posada@hatch.com

References

[1] Posada E. Strategic analysis of alternatives for waste management. In: Kumar S, editor. Waste Management. Rijeka, Croatia: Intech; 2010. ISBN 978-953-7619-84-8

[2] Tello P, Martínez E, Daza D, Soulier M, Terraza H. Informe de la Evaluación Regional del Manejo de Residuos Sólidos Urbanos en América Latina y el Caribe. OPS-AIDIS-BID; 2010

[3] Themelis N, Diaz ME, Estevez P, Gaviota M. Guía para la Recuperación de Energía y Materiales de Residuos. Original study sponsored by Banco Interamericano de Desarrollo 2013. Spanish Edition (Edición en español). Chile/USA: Universidad del Desarollo/ Columbia University; 2016

[4] Min Ambiente-Gobierno de Colombia-Noticias. A 2018 Colombia tendrá una tasa de reciclaje del 20%. 2016. Available from: http://www.minambiente.gov.co/index.php/noticias/2291-a-2018-colombia-tendra-una-tasa-de-reciclaje-del-20

[5] Documento Conpes 3874. Consejo Nacional de Política Económica y Social. República de Colombia. Departamento Nacional de Planeación. Política Nacional Para la Gestión Integral de Residuos Sólidos. 2016. Available from: https://colaboracion.dnp.gov.co/CDT/Conpes/Econ%C3%B3micos/3874.pdf

[6] Global Wtert Council, Research, Education, and Information for Sustainable Waste Management. Summary Report of 2018 Activities of WtERT Partner Organizations; 2019

[7] Saenz G, Posada E. Economical aspects on the implementation of a plant for MWWTP biosolid and coal gasification in comparison with the use of the biosolid for soil recuperation. In: 32nd International Conference on Thermal Treatment Technologies & Hazardous Waste Combustors; 21-23 October 2013; San Antonio, TX. 2013

[8] Saenz G, Posada E. Thermal treatment and energy recovering equipment for waste water and gas emissions from a condensation polymerization plant. Paper # 15. In: 33rd International Conference on Thermal Treatment Technologies & Hazardous Waste Combustors; 13-15 October 2014; Baltimore, MD. 2014

[9] Posada E et al. Simple determination of physical and thermal properties as a basis for the design and analysis of waste to energy combustion equipment. Paper # 24. In: 35th International Conference on Thermal Treatment Technologies & Hazardous Waste Combustors; 4-6 October 2016; Baton Rouge, LA. 2016

[10] Zafar S. Gasification of Municipal Solid Wastes. Bioenergy Consult. 2009. Available from: https://www.researchgate.net/publication/285295261_Gasification_of_municipal_solid_wastes

[11] Arena U. Process and technological aspects of municipal solid waste gasification. A review. Waste Management. 2012;**32**:625-639. Available from: https://www.sciencedirect.com/science/article/pii/S0956053X11004314

[12] Klein A. Gasification: An Alternative Process for Energy Recovery and Disposal of Municipal Solid Wastes; Earth Engineering Center, Colombia University. 2002. Available from: http://www.seas.columbia.edu/earth/wtert/sofos/klein_thesis.pdf

[13] Koukouzas N, Katsiadakis A, Karlopoulos E, Kakaras E. Co-gasification of solid waste and lignite—A case study for Western

Macedonia. Waste Management. 2008;**28**(7):1263-1275. Available from: https://www.sciencedirect.com/science/article/pii/S0196890417309287

[14] Hu B, Huang Q, Buekens A, Chi Y, Yan J. Co-gasification of municipal solid waste with high alkali coal char in a three-stage gasifier. Energy Conversion and Management. 2017;**153**: 473-481. Available from: https://www.sciencedirect.com/science/article/pii/S0196890417309287

[15] Bhoi PR, Huhnke RL, Kumar A, Indrawan N, Thapa S. Co-gasification of municipal solid waste and biomass in a commercial scale downdraft gasifier. Energy. 2018;**163**(15):513-518. Available from: https://www.sciencedirect.com/science/article/abs/pii/S0360544218316876

[16] Eghtedaei R, Mirhosseini SA, Esfahani MJ, Foroughi A, Akbari H. Co-gasification of biomass and municipal solid waste for hydrogen-rich gas production. Energy Sources, Part A: Recovery, Utilization, and Environmental Effects. 2017;**39**(14):1491-1496

[17] Tanigaki N, Manako K, Osada M. Co-gasification of municipal solid waste and material recovery in a large-scale gasification and melting system. Waste Management. 2012;**32**(4):667-675. Available from: https://www.sciencedirect.com/science/article/pii/S0956053X11004788

[18] Available from: https://enerkem.com/

[19] Available from: https://www.mhi.com/products/environment/pyrolysis_gasification_melting.html

[20] Available from: http://www.abengoa.es/web/es/noticias_y_publicaciones/noticias/historico/2017/11_noviembre/20171107_NP_Fulcrum.html

[21] Available from: https://ariescleanenergy.com/case_studies/fluidized-bed-gasification/

[22] Available from: https://ariescleanenergy.com/case_studies/new-tennessee-waste-to-energy-plant/

[23] Available from: https://ariescleanenergy.com/case_studies/city-of-covington-waste-to-energy-gasification-plant/

[24] Available from: https://ariescleanenergy.com/case_studies/replacing-natural-gas-for-industry/

[25] Tanigaki N, Kashiwabara T, Nagata T. Operating Experience from Japanese Waste Gasification Plants with Direct Melting System. Pyrolysis/Gasification. 2017. pp. 379-388

[26] Available from: https://www.globalsyngas.org/resources/map-of-gasification-facilities/

[27] Altmann E, Kellett P. Thermal municipal solid waste gasification. In: Renewable Energy Information Office. Status Report; Irish Energy Center. 1999. Available from: http://www.bluegreenearth.com/site/INCIN/incin/pdf/gasification.pdf

[28] Jones C. Thermal Processing of Waste. 1st ed. Bookboon. Aberdeen: J.C. Jones & Ventus Publishing ApS; 2010. ISBN 978-87-7681-590-5. Available from: http://www.znrfak.ni.ac.rs/serbian/010-studije/oas-3-2/PREDMETI/III%20GODINA/316-KOMUNALNI%20SISTEMI%20I%20ZIVOTNA%20SREDINA/SEMINARSKI%20RADOVI/2014/S151%20-%20S160.pdf

[29] Valkenburg C, Gerber MA, Walton CW, Jones SB, Thompson BL, Stevens DJ. Municipal Solid Waste (MSW) to Liquid Fuels Synthesis, Volume 1: Availability of Feedstock and

Technology. Richland, Washington: Pacific Northwest National Laboratory, U.S. Department of Energy; 2008. Available from: https://www.pnnl.gov/main/publications/external/technical_reports/PNNL-18144.pdf

[30] Tan Y. Feasibility Study on Solid Waste to Energy: Technological Aspects. Berkeley: College of Engineering, Fung Institute, University of California. 2013. Available from: http://funginstitute.berkeley.edu/wp-content/uploads/2014/01/SolidWasteToEnergy.pdf

[31] Gershman, Brickner & Bratton, Inc. Gasification of Non-Recycled Plastics From Municipal Solid Waste In the United States: Thermal Municipal Solid Waste Gasification. American Chemistry Council. 2013. Available from: https://plastics.americanchemistry.com/Sustainability-Recycling/Energy-Recovery/Gasification-of-Non-Recycled-Plastics-from-Municipal-Solid-Waste-in-the-United-States.pdf

[32] Byun Y, Cho M, Hwang S-M, Chung J. Thermal plasma gasification of municipal solid waste (MSW). In: Yun Y, editor. Gasification for Practical Applications. Rijeka, Croatia: IntechOpen; 2012. DOI: 10.5772/48537. Available from: https://www.intechopen.com/books/gasification-for-practical-applications/thermal-plasma-gasification-of-municipal-solid-waste-msw

[33] Rylander H. History of Gasification of Municipal Solid Waste through the eyes. WtERT Blog, posted by Ranjith Annepu. 2012. Available from: http://wtert.blogspot.com/2012/04/history-of-gasification-of-municipal.html

[34] Tangri N, Wilson M. Waste gasification & pyrolysis: High risk, low yield processes for waste management. A Technology Risk Analysis. In: Waste Gasification and Pyrolysis Technology Risk Assessment. GAIA. 2017. Available from: http://www.no-burn.org/gasification-pyrolysis-risk-analysis/

[35] Prepared by Authors Based in Several Studies made During Their Work with Coal Boilers and Power Plants at Colombia

Chapter 2

Syngas Production, Properties, and Its Importance

Raghda Ahmed El-Nagar and Alaa Ali Ghanem

Abstract

Much attention has been focused on reducing the use of petroleum products as fuels, so synthetic gas (Syngas) introduces a great opportunity for energy sustainable developments. Syngas is created either by gasification of plants biomass or waste products (carbon-based) pyrolysis. In principle, Syngas can be produced from any hydrocarbon feedstock. It mainly affects the combustion process in internal combustion engines. The most important is flammability limit, which is very important in the safety and the laminar flame velocity or burning velocity, which is an essential parameter for the investigation of combustion chamber operation and emission performance. This chapter generally reviewed the syngas sources, production, properties, and its importance in the sustainable development for energy.

Keywords: synthetic gas (syngas), petroleum products, pyrolysis, gasification, thermo-chemical process, partially combustion, flammability

1. Introduction

The synthesis gas is defined as a gas with H_2 and CO as the main components of fuel. Row syngas contains mainly significant amounts of CO_2 and H_2O as well. Since syngas is usually used at higher pressures for synthesizing chemicals and fuels (**Figure 1**), the N_2 contents must usually be minimized in syngas. Bio-syngas, however, are biomass produced, chemically identical to syngas. This definition is in accordance with the SYNBIOS-conference definition.

Syngas differ chemically from gasses normally generated by gasification processes at low temperatures, including fluidized bed reactors [1]. To be clear, the gas produced by such reactors is referred to in this report as "product gas." Product gas is defined as a fuel gas with H_2 and CO as well as with substantial amounts of hydrocarbons, such as methane. Product gas contains CO_2 and H_2O, and often N_2, also inevitably.

Throughout the chemical industry, syngas is a substantial intermediate product. Each year, around 6 EJ of syngas are manufactured globally, which is almost 2% of the world's current primary energy consumption. The ammonia industry dominates the global market for syngas (mainly from fossil fuels like coal, natural gas, and oil/residues) [2]. The production of hydrogen for use in refineries, for instance, the processing of hydrogen, and methanol are other major applications. The current market distribution of syngas is shown in **Figure 1**.

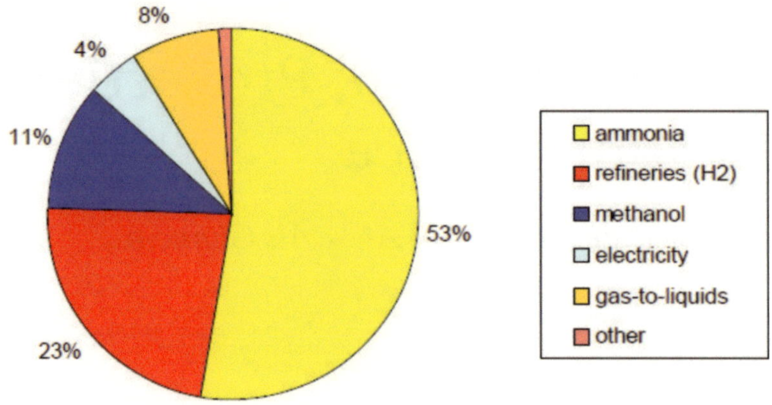

Figure 1.
Present world syngas market, totally ~6 EJ/y.

2. Major properties of syngas

Different characteristics of syngas can affect the process of combustion in internal combustion (IC) motors. The flammability limit of the syngas is one of the most important properties in IC engine safety and fuel. Also, the laminar flame velocity [3] (burning velocity) is an essential parameter to investigate the operation of the combustion chamber and its emission performance.

2.1 Syngas flammability limits

The limit of flammability is usually used as an index for the flammability of the gas. This describes the range of the fuel concentrations in the fuel/air mixture at certain temperature and pressure, which allow the ignition of the flame to propagate and sustain the flammability limits [4] are known in line with generally accepted usages as those fuel-air areas where flame propagation can take place and where fire cannot propagate. The fuel, the spread direction, the size and the form of the combustion chamber, the temperature, and the pressure are primarily affected [4]. And for the fuel-air blend, there are two distinct flammability limits, namely the smallest fuel boundary the flames can propagate is called the lower flammability boundary (LFL), while the richest one is called the upper flammability boundary (UFL). The fact that H_2 and CO are the principal flame-retardant components of syngas inherits the characteristics of these gasses. The presence of inert gasses such as nitrogen and carbon dioxide in the gas mixtures reduces the flammability limit.

2.2 Laminar flame velocity

The laminar flash speed is the speed at which the flame propagates in the direction of expansion wave surfaces under a laminar flow condition via quiet unbranded fuel-oxidant mixes [5]. Because LFV is highly sensitive to combustion chamber operations and emission performance, it is very important for the investigation of combustion chamber operations. The composition of the fuel, mixture equivalence ratio, temperature, and pressure affects it.

2.3 Syngas composition and its calorific value

The composition of the manufacturer's gas depends on feedstock, particulate size, gas flow rate and feedstock flow, chemical reactor configurations, operating conditions or process of gasification, gasificator and catalyst, and gas residence time. But the temperature of the reactor, which in turn is affected by the ER value, mainly influences it. Furthermore, CO, H_2, and CH_4 concentrations in producer gas are also controlled by chemical reactions in the process of gasification.

There is, therefore, a considerable influence on the calorific value of the producer gas on the type of oxidizing agent used for gasification. As ER increases and then the concentrations of these useful components decreases because of the intensification of combustion at higher ER values, the concentrations of CO and H_2 reach the maximum value. As ER increases, the concentrations of CO_2 and N_2 in the producer gas are also increased [4]. Air as an oxidant produces syngas with relatively high levels of nitrogen and thus a lower heat value, which does not normally exceed 6 MJ/Nm^3.

The producer's gas is classified as fuel gas of low quality. The typical biomass gasification composition of an air-borne downdraft reactor with the oxidizer is as follows: 15–20% of H_2, 15–20% of CO, 0.5–2% of CH_4, 10–15% of CO_2, and the rest of the component of N_2, O_2, and CXHY. If the concentration of fuel components is considerately increased and the gas is called a medium heat value, up to 16 MJ/Nm^3 [6], where oxygen or water steam or the mixture of both are used.

3. Production of synthesis gas

Synthesis gas or syngas is called carbon monoxide (CO)-hydrogen (H_2)-containing gas mixture. Carbon dioxide (CO_2) and other components such as water (H_2O) may also be present in syngas.

The chemical synthesis can be used as a building block in all products normally produced from crude oil or natural gas. Petrol and diesel are fuels with definitions, which are not based on the chemistry but on their physical qualities such as boiling and flashing.

The octane rating of gasoline in an internal combustion (IC) engine is empirical and based on its actual performance. This means that at least low-grade blending is possible as long as synthetic fuel matches the characteristics of petrol- or diesel-based crude oil.

3.1 Production of fuels and chemicals from gasification of biomass/coal or reforming of natural gas

The feedstock must be gasified when beginning with a solid feedstock, such as biomass or coal. It may be necessary to ground or pulverize the feedstock before gasification (usually carbon). The particle fineness depends on the gasification type. For most biomass gasification plants, drying is required as the next step. The drying is integrated into a gasifier reactor vessel by certain gasifiers [7].

Following gasification, the product is a gas, known as producer gas, filled with impurities that must be removed. Prior to synthesis, the producer of gas usually also needs the ratio from H_2 to CO. H_2:CO Both.

3.2 Gasification

All hydrocarbon feed resources like coal, heavy oils, or combustible biomass can be gasified as synthesis gas. Several reactions occur in the gasificator, but the total reaction can be summarized by Eqs. (1) and (2).

$$\text{Biomass} + O_2 \rightarrow CO + H_2 + CO_2 + H_2O + CH_4 \tag{1}$$

Composition	Updraft Gasifier (% by Volume)	Downdraft Gasifier (% by Volume)
Carbon monoxide	24	21
Hydrogen	11	17
Methane	3	2
Hydrocarbon	0.2	0.3
Nitrogen	53	48
Water vapor	3	4

Table 1.
Composition of producer gas [8].

By-products: tar, char, ashes.
Reaction conditions:

- High temperature (800–1000°C) and low pressure (1–20 bar)
- H_2/CO ratio within 0.5 and 1.8, depending on the technologies

The reaction also can be expressed as:

$$CaHbOcNd + O_2/H_2O/N_2 \rightarrow CO + H_2 + CxHyOz + CO_2 + H_2O + NH_3 + N_2 \quad (2)$$

The solid carbon is partially oxidized with oxygen (O_2), air, steam (H_2O), or a combination of all gasification agents. CxHyOz is mostly made of methane with a few low percent of hydrocarbon, including ethane and ethylene (**Table 1**). For most gasifiers, gas may also contain heavier hydrocarbons such as benzene, toluene, and naphthalenes, depending on the feedstock and the operational parameters. Hydrocarbons, which are heavier than benzene, are often known as tars (**Figure 2**).

This is an important factor in determining the technical mechanism and the economic feasibility of the gasification system. Efficiencies in gasification are based on the biomass type used, its particle size, its ER value, and the reactor design [8].

The gasification efficiency is usually determined on the lower heating value basis. The efficiency is calculated as the ratio of the total energy in the producer gas (sensible and chemical) and the chemical energy in the feedstock (the heating value). Depending on type and design of the gasifier as well as on the characteristics of the fuel, mechanical gasifier efficiency may vary between 60 and 75%. A useful definition of the gasification efficiency (%) used for engine applications is as follows:

$$\eta_m = \frac{H_g \times Q_g}{H_s \times M_s} \times 100$$

where ηm is the gasification efficiency (%) (mechanical), Hg is the heating value of the gas (kJ/m^3), Qg is the volume flow of gas (m^3/s), Hs is the lower heating value of gasifier fuel (kJ/kg), and Ms is the gasifier solid fuel consumption (kg/s).

3.3 The yield of syngas

The syngas yield is measured by the mass of the produced in cubic meters per the mass of the feedstock provided to the system The yield is directly commensurate with the difference in ER and with the gas residence time in the reduction area [9]. The biomass ash content also has a considerable impact and limits the yield of the gas producer.

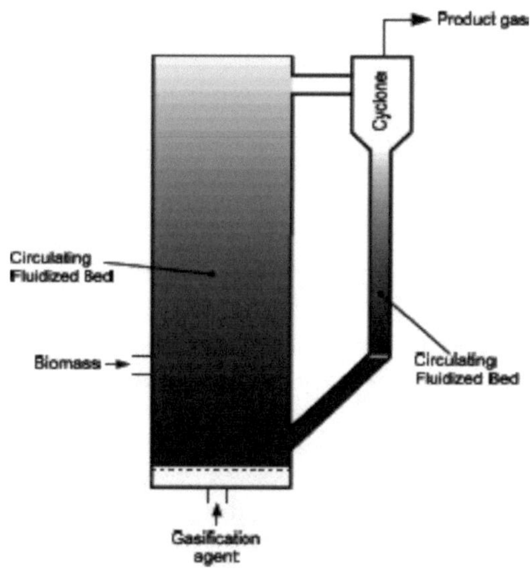

Figure 2.
The gasificator [8].

3.4 Cleaning and cooling of producer gas

The combustible gas can be used as a feedstock for the production of chemicals like methanol or in internal combustion engines for direct heat uses.

4. Fischer-Tropsch diesel production

One of the commercially accessible methods of manufacturing clean synthetic fuel from syngas is Fischer-Tropsch technique. Industrially speaking, coal/pet-coke/biomass emits huge quantities of carbon dioxide that can be used to enhance the fuel production.

Because of the increased level of carbon dioxide in atmosphere and also the depletion of conventional fuels, scientific researches recommend the chemical recycling of carbon dioxide into renewable fuel and more added-value chemicals [8].

The production of synthetic fuel from syngas is favorable process owing to its portability as well as to its large quantity of chemical energy, saved without further processing and easy to use.

The raw syngas undergoes multiple energy intensive processes to fulfill the stoichiometric requirements ($2.05 < H_2/CO < 2.15$) for Fischer-Tropsch, including removal of carbon dioxide for subsequent sequestrating, to mitigate negative carbon dioxide emission impacts. Absorption techniques with mono- and diethyl amines are the most popular technology for the removal of carbonic gases from syngas. These procedures are energy intensive owing to the intermediate steps of absorption, desorption, and compression.

The main Fischer-Tropsch reaction is to produce aliphatic long-chain saturated hydrocarbons from syngas (Eq. (3)). There are a lot of side reactions that occur on active sites accompanying the main reaction (Eq. (4)).

Original FT reaction:

$$nCO + 2nH_2 \rightarrow (CH_2)n + nH_2O; \Delta H298\ K = -152\ kJmol^{-1} \quad (3)$$

Carbon dioxide is also a Fischer-Tropsch waste product and considered to effect the targeted yield of the produced liquid hydrocarbons, and also the presence of carbon dioxide can lead to a substantial reduction in the catalytic reaction activity [10]. This is due to the water-gas-shift (WGS) reaction (Eq. (4)) by which the hydrogen lack is overcome and releases a large quantity of carbon dioxide during the reaction.

Water-gas-shift reaction:

$$CO + H_2O \rightleftharpoons CO_2 + H_2; \Delta H298\ K = -41.2\ kJmol^{-1} \tag{4}$$

It was reported that the carbon dioxide found in syngas plays the role of an oxidizing agent on reduced $Co/\gamma\text{-}Al_2O_3$, which effects the conversion of carbon monoxide and C5+ hydrocarbons selectivity. Another role for the carbon dioxide was proposed as an inert gas in cobalt-based catalysts [10]. Another approach described the formation of dioxide during the Fischer-Tropsch reaction that will reduce with the using of syngas-containing carbon dioxide as the equilibrium tends to be directed in the inverse direction without affecting the Fischer-Tropsch process.

The change in equilibrium of carbon dioxide is the first step in Fischer-Tropsch (Eq. (6)), resulting in an enhanced proportion of oxygen atoms in carbon monoxide, which is retrieved by water over iron-based carboxylic catalysts comprising syngas. The produced carbon monoxide from this shift is further processed in the Fischer-Tropsch technique.

Boudouard reaction:

$$2CO \rightleftharpoons CO_2 + C(s); \Delta H298\ K = -172.5\ kJmol^{-1} \tag{5}$$

Modified CO_2-FT reaction:

$$CO_2 + H_2 \rightleftharpoons CO \rightarrow 2nH_2 \rightarrow (CH_2)n + nH_2O + H_2O \tag{6}$$

Direct CO_2 hydrogenation:

$$nCO_2 + 2nH_2 \rightarrow (CnH_2n)n = 2\text{–}4 + nH_2O \tag{7}$$

The low selectiveness of 10Co5Fe supported on carbon nanofiber catalyst to produce methane during Fischer-Tropsch reaction. The increase in carbon dioxide levels in syngas produces only 22% of C5+ hydrocarbons.

Fe_3O_4 catalyzes the transformation of carbon dioxide into carbon monoxide through the inverse response of the water-gas-shift system, while $\chi\text{-}Fe_5C_2$ is involved in hydrocarbon production.

Catalytic transformation of syngas was demonstrated with the use of bifunctional Fe-Co, backed on hierarchical HZSM-5, of 16% (mol%) carbon dioxide in hydrogen deficiencies. 1Fe:2Co (wt%) is the most effective bimetallic mixture of several combinations of iron and cobalt.

5. Conclusion

- The synthesis gas is defined as a gas with H_2 and CO as the main components of fuel.
- The flammability limit of the syngas and the laminar flame velocity are the major syngas properties.

- Syngas can be produced from gasification of biomass/coal or reforming of natural gas, and the yield is measured by the mass of the produced in cubic meters per the mass of the feedstock.

- Fischer-Tropsch technique is one of the commercially accessible methods of manufacturing clean synthetic fuel from syngas.

Author details

Raghda Ahmed El-Nagar[1*] and Alaa Ali Ghanem[2]

1 Analysis and Evaluation Department, Egyptian Petroleum Research Institute (EPRI), Egypt

2 Production Department, Egyptian Petroleum Research Institute (EPRI), Egypt

*Address all correspondence to: raghda_elnagar@yahoo.com

References

[1] Abdelkader A, Daly H, Saih Y, Morgan K, Mohamed MA, Halawy SA, et al. Steam reforming of ethanol over Co_3O_4-Fe_2O_3 mixed oxides. International Journal of Hydrogen Energy. 2013;**38**:8263-8275

[2] Díaz JA, Akhavan H, Romero A, Garcia-Minguillan AM, Romero R, Giroir-Fendler A, et al. Cobalt and iron supported on carbon nanofibers as catalysts for Fischer-Tropsch synthesis. Fuel Processing Technology. 2014;**128**:417-424

[3] Esmaili E, Mostafavi E, Mahinpey N. Economic assessment of integrated coal gasification combined cycle with sorbent CO_2 capture. Applied Energy. 2016;**169**:341-352

[4] Jiang F, Zhang M, Liu B, Xu Y, Liu X. Insights into the influence of support and potassium or sulfur promoter on iron-based Fischer–Tropsch synthesis: Understanding the control of catalytic activity, selectivity to lower olefins, and catalyst deactivation. Catalysis Science & Technology. 2017;7:1245-1265

[5] Mandal S, Maity S, Gupta PK, Mahato A, Bhanja P, Sahu G. Synthesis of middle distillate through low temperature Fischer-Tropsch (LTFT) reaction over mesoporous SDA supported cobalt catalysts using syngas equivalent to coal gasification. Applied Catalysis A: General. 2018;**557**:55-63

[6] Shashank B, Anna MV, Sreedevi U. Utilization of greenhouse gas carbon dioxide for cleaner Fischer-Tropsch diesel production. Journal of Cleaner Production. 2019;**228**:1013-1024

[7] Yu X, Zhang J, Wang X, Ma Q, Gao X, Xia H, et al. Fischer-Tropsch synthesis over methyl modified Fe_2O_3@ SiO_2 catalysts with low CO_2 selectivity. Applied Catalysis B: Environmental. 2018;**232**:420-428

[8] Visconti CG, Martinelli M, Falbo L, Infantes-Molina A, Lietti L, Forzatti P, et al. CO_2 hydrogenation to lower olefins on a high surface area K-promoted bulk Fe-catalyst. Applied Catalysis B: Environmental. 2017;**200**:530-542

[9] Halder A, Kilianová M, Yang B, Tyo EC, Seifert S, Prucek R, et al. Highly efficient Cu-decorated iron oxide nanocatalyst for low pressure CO_2 conversion. Applied Catalysis B: Environmental. 2018;**225**:128-138

[10] Kordouli E, Pawelec B, Kordulis C, Lycourghiotis A, Fierro JLG. Hydrodeoxygenation of phenol on bifunctional Ni-based catalysts: Effects of Mo promotion and support. Applied Catalysis B: Environmental. 2018;**238**:147-160

Chapter 3

Sustainability Effect of Water Gas Shift Reaction (Syngas) in Catalytic Upgrading of Heavy Crude Oil and Bitumen

Onoriode P. Avbenake, Faruk Ibrahim Yakasai and Baba Y. Jibril

Abstract

Industrial globalization and urbanization has placed man on a path for the quest of energy. Conspicuously there are various types and forms of energy with established technologies on harnessing and utilization wherein fossil fuel still remains the cheapest with straightforward application. Although, conventional fossil fuel resources are depleting because of population growth, unconventional reservoirs are exploited daily with limited production due to inadequate and expensive technology making them stand at approximately 9.1 trillion barrels poised to quell the world's energy insecurity for the next 100 years. The underlying basis for the high-cost of unconventional reservoirs upgrading is the hydrogen required to seal the alkyl chain after cracking to form low molecular weight hydrocarbons. Therefore, in this chapter we aim to provide a comprehensive review of in-situ generation of hydrogen from syngas via water gas shift reaction for the catalytic upgrading of heavy crude oil and bitumen by analyzing the gas chromatography results of gaseous effluents for the presence of syngas in the various works cited. Although, heavy crude oil and bitumen are non-renewable the upgrading method selected is tenable, appropriately the overall technology is partially sustainable.

Keywords: fuel, syngas, water gas shift reaction, heavy crude oil, bitumen, hydrogen, gas chromatography, sustainability

1. Introduction

Synthesis gas (Syngas) is a gas mixture containing carbon monoxide (CO) and hydrogen (H_2) in dissimilar proportions produced from gasification of a carbon-containing material to gaseous products [1]. Also, water gas reaction (WGR) is a mixture of carbon monoxide and hydrogen produced by passing steam over red-hot coke in an endothermic reaction (Eq. 1).

$$C + H_2O \rightarrow CO + H_2 \qquad (1)$$

$$C + CO_2 \rightarrow 2CO \qquad (2)$$

$$CO + H_2O \leftrightarrow CO_2 + H_2 \qquad (3)$$

Reaction 3 in the above mechanism is foremost in order to shift the carbon monoxide entirely to hydrogen in the presence of an oxide based catalyst. This reaction can be applied separately and it is called the water gas shift reaction (WGSR) [2]. Detailed reaction mechanism for syngas production is displayed in **Table 1** to depict where all these fit in.

Reactions in syngas production follows series of steps subdivided into; reforming, oxidation and carbon forming with water gas shift passing off as a very rapid equilibrium reaction step. These three distinct reactions are independent in their own right and characterized by the reaction condition or active ingredient. Reforming reaction is partial combustion of methane or other hydrocarbon sources in the presence of water to form carbon monoxide and hydrogen, while oxidation reaction is the partial combustion of methane in oxygen to give carbon monoxide, carbon dioxide and hydrogen. On the other hand, carbon forming reaction is the reversible pyrolysis of methane or other hydrocarbon sources and disproportionation of carbon monoxide to carbon dioxide to form carbon and syngas.

Moreover, reports of current depletion of conventional fossil fuel reservoirs [3, 4] and increasing discoveries of heavy crude oil and bitumen deposits around the world [5, 6] has threatened global industrialization. However, this energy insecurity perceived in the near future could be averted with unconventional reservoirs upgrading which currently stands at 9.1 trillion barrels making it 70% of the world total oil resources [5].

Conversely, syngas is a promising fuel [7]. In the past it has been used to provide hydrogen for various industrial applications [8], and fuel sources in the case of fuel cells [9] and methanol synthesis [10]. In Fischer-Tropsch synthesis, it is used to manufacture liquid fuels from gas, coal or biomass [11]. Recently, it has found application as a direct fuel in hydrogen internal combustion engines for land and air transportations [12].

In this chapter we attempt to review the sustainable application of syngas in heavy crude oil and bitumen upgrading via water gas shift reaction. Various works on this area were critically discussed to ascertain the participation of syngas in the upgrading process via gas chromatography analysis of the gaseous product which shows high proportions of carbon dioxide and hydrogen when the starting materials are simply water in a neutral environment with a suitable catalyst and the heavy crude oil/bitumen hydrocarbon. Although, these hydrocarbons are non-renewable the upgrading method selected in this chapter is tenable, as a consequence the overall technology is partially sustainable.

Reforming reactions $CH_4 + H_2O \leftrightarrow CO + 3H_2$ $C_nH_m + mH_2O \leftrightarrow nCO + (n + ½m)H_2$	+206 kJ/mol Endothermic	Steam methane reforming (SMR)
Oxidation reactions $CH_4 + ½O_2 \rightarrow CO + 2H_2$ $CH_4 + O_2 \rightarrow CO + H_2O + H_2$ $CH_4 + 3/2O_2 \rightarrow CO + 2H_2O$ $CH_4 + 2O_2 \rightarrow CO_2 + 2H_2O$	−36 kJ/mol −278 kJ/mol −519 kJ/mol −802 kJ/mol	Partial combustion
Carbon forming reactions $CH_4 \leftrightarrow C + 2H_2$ $2CO \leftrightarrow C + CO_2$ $CO + H_2 \leftrightarrow C + H_2O$ $C_nH_m \leftrightarrow nC + ½m\ H_2$	+75 kJ/mol −172 kJ/mol −131 kJ/mol Endothermic	Boudouard reaction

Table 1.
Reactions in syngas production.

2. Heavy oil upgrading

Heavy crude oil and bitumen upgrading technologies can be classified into; carbon rejection, hydrogen addition (hydrocracking) and separation processes [13, 14]. Hydrocracking is a process of upgrading heavy crude oil in the presence of hydrogen and a suitable catalyst, whereby the latter is usually dual functional with the hydrogenating and cracking sites, while the former inhibits secondary reactions that produce coke [15]. However, hydrocracking as simply defined above has been established to be the most suitable technology for heavy crude oil and bitumen upgrading [16]. Consequently, various methods abound for hydrocracking processes with distinct strategy along the lines of temperature and pressure operating conditions, type of reactors, composition and type of catalysts, and ultimately the methods of providing hydrogen.

Hydrogen can be supplied directly as a pure gas [17–20], produced in-situ from chemical compounds and biological species [21–24], or extracted from syngas process in-situ via water gas shift reaction [20, 25–28]. In the first case, a significant amount of hydrogen proceed to hydrogen sulfide (H_2S) [18, 20] and the process is over the odds [19]. The impediment of the second method is low conversion [22] and inability to operate at relatively higher reaction conditions due to biological species [23, 24]. However, supply of hydrogen via syngas is inexpensive [25], has high conversion rate [26–28] and the reaction could be operated between low water gas shift reaction (200–350°C) [28, 29] and high water gas shift reaction conditions (350–450°C) [20, 27]. In addition, higher upgrading has been reported with processes where hydrogen was obtained via syngas to pure hydrogen gas [26].

2.1 Catalytic reaction in neutral environment

In heavy crude oil and bitumen upgrading where the hydrogen is supplied from syngas, the source is usually from water in-situ [13] or introduced into the reactor alongside feedstock [20, 25–31]. Irrespective of the method of providing water, the reactor would be pressurized in a neutral environment, usually nitrogen to mimic reservoir conditions.

Fumoto et al. [32] studied the suppression of coke generation in upgrading of bitumen by examining the time factor (W/F) of the catalyst and mixture of steam and nitrogen as feedstock at 500°C and atmospheric pressure. Meanwhile, Chao et al. [33] developed a new type of difunctional catalyst in heavy oil upgrading. The starting materials include; 100 g of heavy crude oil and water in a designed mass ratio, catalyst and nitrogen gas at 240°C and 3 MPa. Similarly, the role of water in the redox reaction between bitumen and water in the presence of a suitable catalyst was studied by Dejhosseini et al. [25]. They observed that bitumen cracking was supported either through oxidation of active oxygen species generated from water or hydrogen via the redox (syngas) reaction (**Figure 1**).

2.2 Production of syngas

Syngas process follows a series of endothermic and exothermic reaction steps, subdivided into three major units (not in order of occurrence); reforming, oxidation and carbon forming reactions as shown in **Table 1** [34].

It is suitable to state here that only at temperatures above 700°C would the SMR dominate carbon formation reactions, also the other steps of syngas production are reversible reactions except oxidation reactions which only goes in the forward direction.

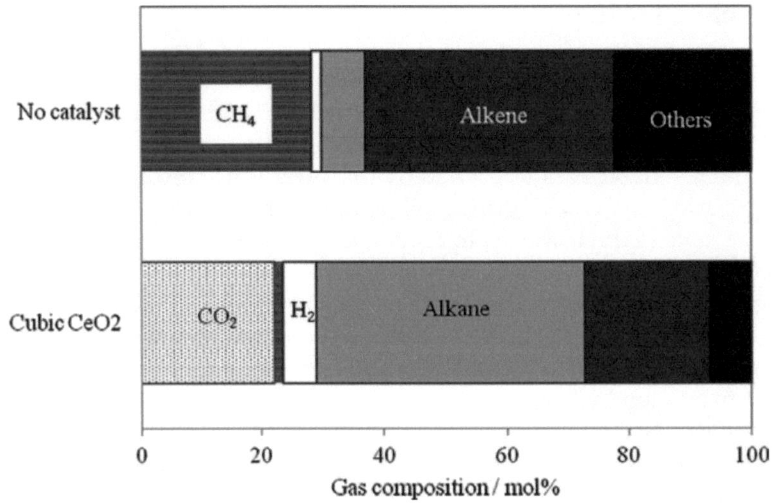

Figure 1.
Gas composition after the reaction of bitumen in presence and absence of cubic CeO_2 nanoparticles at 450°C in supercritical water [25].

To ascertain the process of syngas during hydrocracking reactions, researchers usually utilize gas chromatography to analyze the produced gas. Results obtained show that in the presence of water and a suitable catalyst, these gases contain a considerable amount of carbon dioxide, carbon monoxide and low quantity of hydrogen assuming majority have been consumed by free alkyl chain molecules to form light oils [20, 25].

Certainly, heavy crude oil or bitumen reactivity and hydrogen availability are the two major factors responsible for the extent of upgrading [35]. Accordingly, the latter is the bedrock of upcoming technologies since the former is inflexible. In other respect, syngas could directly be injected into the upgrading process with the intention that on contacting carbon monoxide with the catalyst additional hydrogen would be generated for the upgrading reaction [36]. Bitumen can also be upgraded in supercritical water, hydrogen and carbon dioxide mixtures with the hydrogen and carbon dioxide species produced via decomposition of a carboxylic acid in the presence of water [37].

GC-MS analysis of the produced gas after upgrading with and without catalysts in the presence of water was used to measure the extent of upgrading and participation of water gas shift reaction in the overall process. **Figure 2** displays the GC-MS chromatogram of upgrading of heavy crude oil in the presence of alkyl ester sulfonate copper catalyst and water, where (a) is without catalysts and (b) with catalysts [33]. The first peak is carbon dioxide with abundance almost twice as much with catalysts than without catalysts. Since carbon dioxide is a product of water gas shift reaction alongside hydrogen in 1:1 stoichiometric molar ratio, it is reasonable to conclude that syngas is involved in the catalytic upgrading of heavy oil in the presence of water actively reducing the viscosity by 90%.

Fumoto et al. [38] exploit syngas in the form of water gas shift reaction for oxidative cracking of residual oil to produce useful light hydrocarbon fuels. Analysis of the gaseous product of the cracking process depicts carbon dioxide predominates. In the experiment without catalysts, methane and other lower saturates and alkenes constituted the gaseous effluents. Active hydrogen and oxygen species were formed from water via syngas which are consumed in the process thereby producing excess carbon dioxide and hydrogen as shown in **Figure 3**. In a similar manner, Ajumobi et al. [39] analyzed the gaseous effluents of the upgrading of oil sand bitumen via

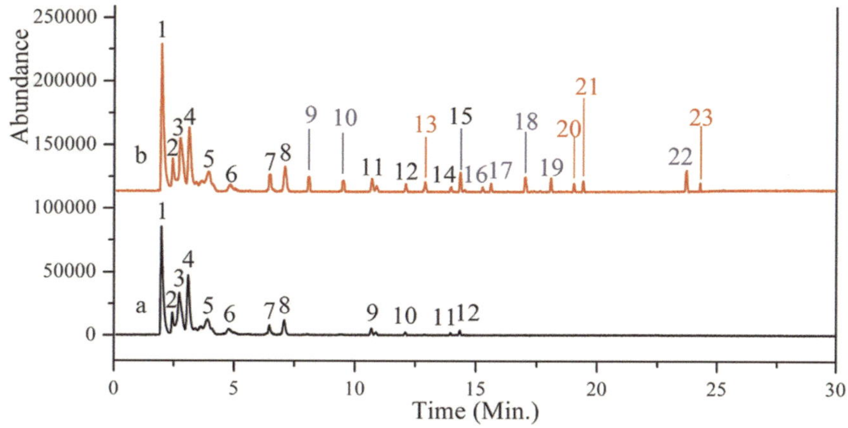

Figure 2.
GC-MS spectra of the pyrolytic gas after reaction (a) without and (b) with the difunctional catalyst [33].

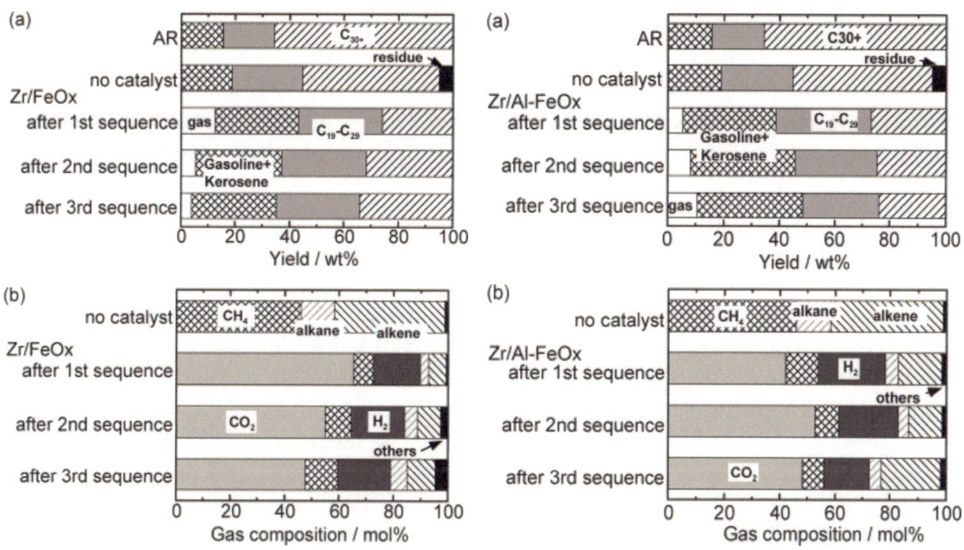

Figure 3.
(a) Liquid product yield and (b) gas compositions after the sequence of the reaction of residual oils with steam over Zr/FeO$_x$ and Zr/Al-FeO$_x$ catalyst and regeneration [38].

ceria-based catalysts in steam environment and reported higher compositions of carbon dioxide and hydrogen in the experiments with catalysts as against that without catalyst resulting in higher light oil yields of approximately 60 mol% carbon. Therefore, it is evident from the foregoing that without the presence of a suitable catalyst to provide active hydrogen from water components the upgrading process would be impaired. Hosseinpour et al. [40] observed that the maltenes transforms to coke during vacuum residue upgrading in supercritical water.

3. Application of syngas

Syngas dates back to energy but over the years goes by different names depending on its starting material; producer gas, town gas, blue water gas, water gas, synthesis gas, syngas and so on. Earlier in this chapter it has been shown that water gas shift reaction is a pertinent step in syngas production.

Owing to its significance and age, extensive research has been dedicated to the study of syngas ranging from production to application. On the whole, syngas can be produced from gas, oil, coal or biomass via steam reforming, partial oxidation or gasification. The as-synthesized syngas can be applied in Fischer-Tropsch synthesis to produce synfuels, methanol synthesis to produce methanol or water gas shift reaction to produce hydrogen.

3.1 Syngas to synfuels

Liquid fuels can be manufactured from syngas with one of three processes; gas-to-liquids (GTL), coal-to-liquids (CTL), and biomass-to-liquids (BTL) via the Fischer-Tropsch synthesis [41]. The synthesis was developed in 1923 as a series of catalyzed chemical reactions which could be summarized to the basic reaction (Eq. 4),

$$nCO_{(g)} + [n + m/2]H_{2(g)} \rightarrow C_nH_m + nH_2O_{(g)} \quad (4)$$

with the liquid fuel hydrocarbon building from the methylene group attaching sequentially on a carbon chain. Although there have been extensive work on the Fischer-Tropsch synthesis, the reaction mechanism is still vague. Generally, the process requires a cobalt or iron catalyst [42, 43] to convert any carbonaceous raw material into synthetic natural gas, liquefied petroleum gas, light and heavy petroleum, middle distillate, kerosene, Diesel, and waxes [44, 45].

3.2 Syngas to methanol

Methanol is an important industrial intermediate and doubles as a medium for hydrogen storage. The synthesis is as old as wood fires, as history tells that the methanol used in World War I was derived from destructive distillation of wood [46]. Unlike the Fischer-Tropsch synthesis, there is established reaction mechanism for catalytic production of methanol from syngas as far back as 1977, later modified by the same authors in 1991 [47]. In their scheme they considered the formation of surface species on the catalyst and production of ethanol as by-product. They further suggested that the chain addition of methylene group to the surface adsorbed aldehydes (species I–IV) are important steps as shown in **Figure 4**. In 2003, Reubroycharoen et al. [48] proposed a more simplified mechanism with Cu-based oxide catalyst using alcohol as catalytic solvent. They successfully produced methanol from syngas with a semi-batch autoclave reactor operating at 170°C and 50 bar reporting a one-pass yield of 47% and a selectivity of 98.9%.

3.3 Syngas to hydrogen

The importance and various application of hydrogen cannot be overemphasized. Hydrogen has found application as industrial chemical in manufacture of fertilizers [41], metal work [44], glass and steel production [46], and coolant in electricity generation [47]. However, its large scale application is still in the energy sector [49] where it is applied directly or indirectly. Direct application of hydrogen is in internal combustion engines [50] and fuel cells [51]. Indirectly hydrogen is used to produce liquid fuel [52], methanol [53], and upgrading of heavy crude oil and bitumen [54].

The best method of heavy crude oil and bitumen upgrading in line with sustainability is in-situ generation of hydrogen considering economic and environmental footprint [55–57]. Applications of syngas by way of water gas shift reaction have

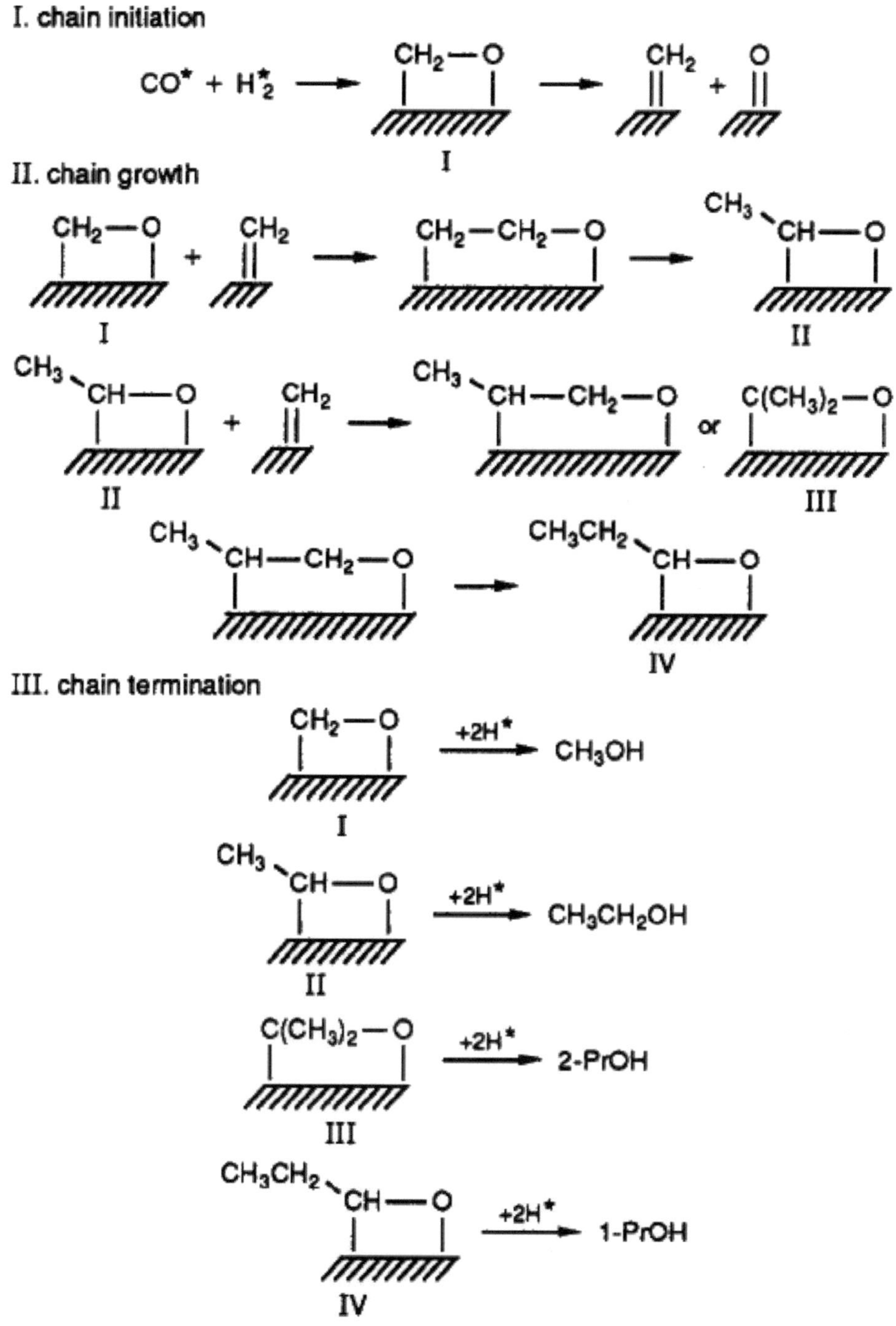

Figure 4.
Reaction steps for methanol synthesis from syngas [47].

produced tremendous results in heavy oil upgrading. In the first place, sufficient reaction mechanisms have been used to explain the process [58, 59], also successful field tests have been conducted [60], and lastly there are ample reports of obtaining better upgrading when hydrogen is supplied via water gas shift reaction as against pure hydrogen gas [13, 27, 61].

Kapadia et al. [62] studied the in-situ generation of hydrogen from bitumen gasification in the presence of methane via water gas shift reaction. They included in their scheme; thermal cracking, low and high temperature oxidation reactions, coke gasification, water gas shift reaction, methanation, and hydrogen and carbon monoxide combustion reactions and solved the following elementary steps Eqs. (5)–(17) derived from various literatures [62, 63];

Thermal cracking reactions

$$Maltenes \rightarrow 0.372164\ Asphaltenes \tag{5}$$

$$Asphaltenes \rightarrow 83.23\ Coke \tag{6}$$

$$Asphaltenes \rightarrow 25.2965\ C_xH_yO_z \tag{7}$$

Low temperature oxidation reactions

$$Maltenes + 0.43O_2 \rightarrow 0.4726\ Asphaltenes \tag{8}$$

$$Asphaltenes + 7.51275O \rightarrow 101.539\ Coke \tag{9}$$

High temperature coke oxidation reaction

$$Coke + 1.232O_2 \rightarrow 0.8995CO_2 + 0.1CO + 0.564H_2O \tag{10}$$

High temperature gas oxidation reactions

$$CH_4 + 2O_2 \rightarrow CO_2 + 2H_2O \tag{11}$$

$$C_xH_yO_z + 2O_2 \rightarrow 0.9695CO_2 + 2H_2O \tag{12}$$

Hydrogen generation reactions

$$Coke + H_2O \leftrightarrow CO + H_2 \tag{13}$$

$$CO + H_2O \leftrightarrow CO_2 + H_2 \tag{14}$$

Hydrogen consumption reactions

$$Coke + 2H_2 \leftrightarrow CH_4 \tag{15}$$

$$H_2 + 0.5O_2 \rightarrow H_2O \tag{16}$$

Carbon monoxide combustion reaction

$$CO + 0.5O_2 \rightarrow CO_2 \tag{17}$$

and successfully predicted hydrogen production, bitumen pyrolysis and aquathermolysis when compared with experimental data. They observed that at higher temperatures the mechanism favors coke gasification and water gas shift reaction resulting in high concentration of carbonaceous gases and ultimately hydrogen.

A comparable experimental investigation was made in the upgrading of heavy crude oil using Co-Mo/Al_2O_3 catalysts. On the one hand steam was supplied as feedstock and on the other hand pure nitrogen was used to pressurize the reactor to observe the effect of steam and ultimately water gas shift reaction [64]. Their results showed that 92% viscosity reduction was obtained in the steam environment as against 85% in nitrogen. The coke content of the product was also decreased to 11 wt% in steam as against 27 wt% obtained with nitrogen. This goes on to show that water gas shift reaction plays an important role in the catalytic upgrading of heavy crude oil to produce required hydrogen needed to prevent the secondary reaction of the carbon-carbon bond scission products during cracking to coke. This hydrogen also come in play to reduce the recombination of active chains formed by cleavage of C–C, C–S, C–N, and C–O bonds to form large molecules [65]. They deduce a simple reaction steps to depict the effect of water Eqs (18)–(20).

$$heavy\ oil + H_2O \rightarrow Light\ hydrocarbon + gases\ (C_1 - C_5, H_2, CO, etc.) + active\ chains \tag{18}$$

$$active\ chains + active\ H_2 \rightarrow low\ molecular\ weight\ compound \tag{19}$$

$$active\ chains + active\ chains \rightarrow high\ molecular\ weight\ compound \tag{20}$$

An Agilent gas chromatograph refinery gas analyzer was used to determine the composition of the produced gases from the nitrogen feedstock and those of varying steam-oil ratio (SOR) as shown in **Table 2**. It could be observed that the total gas produced with nitrogen reaction 6.61142 vol% is twice that produced in the various SOR reactions 3.61152, 3.00644 and 3.64143 vol%. It is learnt that during heavy crude oil cracking, hydrogen is formed from subtraction of excited saturated to unsaturated hydrocarbons and C–H bond cleavage as shown in Eqs. (21)–(23).

$$2RH_{saturated} \rightarrow 2R_{unsaturated} + H_2 \tag{21}$$

$$RH \rightarrow R' + H' \tag{22}$$

$$RH + H' \rightarrow R' + H_2 \tag{23}$$

Gas	No steam (vol%)	SOR (0.1) (vol%)	SOR (0.05) (vol%)	SOR (0.02) (vol%)
Methane	2.33	1.15	1.27	1.46
Hydrogen	2.4	1.37	1.1	1.41
Ethane	0.82	0.06	0.033	0.03
Carbon dioxide	0.03	0.07	0.064	0.054
Ethene	0.03	0.07	0.08	0.073
Propane	0.23	0.21	0.15	0.113
Propene	0.14	0.10	0.07	0.068
i-Butane	0.12	0.14	0.03	0.062
n-Butane	0.06	0.14	0.03	0.073
i-Butene	0.051	0.035	0.034	0.044
cis-2-Butene	0.00	0.037	0.029	0.038
trans-2-Butene	0.02	0.033	0.00	0.023
n-Pentane	0.19	0.11	0.064	0.11
i-Pentane	0.17	0.07	0.034	0.064
Carbon monoxide	0.02	0.016	0.018	0.019
Hydrogen sulfide	0.00042	0.00052	0.00044	0.00043
Total	6.61142	3.61152	3.00644	3.64143

Table 2.
Produced gas composition during catalytic upgrading reaction in nitrogen atmosphere only and in combination with steam [64].

With the steam reaction having more advantage of generating extra hydrogen via water gas shift reaction, it is unexpected to record lower hydrogen composition compared to nitrogen environment. However, it is easy to perceive that the unaccounted hydrogen for the steam experiments have been consumed during hydrogenation, hydrodesulfurization and de-coking as reported earlier. On the contrary, one would expect that the high composition of hydrogen reported in the experiment without steam would increase the upgrading, lower coke formation or improve olefin saturation. Nevertheless, it has been reported that the proton formed via water gas shift reaction has higher hydrogenating reactivity than others [13].

Another observation in **Table 2** is the high volume of methane recorded in the reaction without steam. In the presence of water and a suitable catalyst, methane undergoes steam-methane reforming (SMR) to produce syngas according to Eq. (24).

$$CH_4 + H_2O \rightarrow 3H_2 + CO \quad (24)$$

In a bid to further explore the resourcefulness of syngas in hydrocracking of heavy crude oil and bitumen, hydrogenation could progress either through the forward water gas shift reaction (Eq. 3) or reverse water gas shift reaction (Eq. 25).

$$CO_2 + H_2 \rightarrow CO + H_2O \quad (25)$$

A comprehensive research structure was used to demonstrate this in hydrodesulfurization of dibenzothiophene using NiMo/Al_2O_3 catalysts at 673 K and 30 MPa [61]. Four different mixtures; hydrogen-water, carbon monoxide-water, carbon dioxide-hydrogen-water, and HCOOH-water were used as hydrogen sources, and conversion were obtained in the order; carbon monoxide-water, carbon dioxide-hydrogen-water, HCOOH-water and hydrogen-water. It would be observed that the carbon monoxide-water and carbon dioxide-hydrogen-water combinations are the forward and reverse water gas shift reactions respectively. Product analysis of the carbon dioxide-hydrogen-water mixture revealed trace amount of carbon monoxide a product of reverse water gas shift reaction (Eq. 25).

Basically, there are two mechanisms postulated for the water gas shift reaction; the regenerative mechanism and the associative mechanism. In the regenerative mechanism, the redox reaction on the surface of the catalysts is responsible for the hydrogen production [66]. It is proposed that the catalysts surface is oxidized by water to produce hydrogen followed by reduction of the surface to convert carbon monoxide to carbon dioxide as seen in Eqs. (26) and (27).

$$H_2O + red \rightarrow H_2 + ox \quad (26)$$

$$CO + ox \rightarrow CO_2 + red \quad (27)$$

For the associative mechanism, an adsorption-desorption model was proposed which involves intermediate and eventual desorption to carbon dioxide and hydrogen (see Eq. 28).

$$CO + H_2O \rightarrow (intermediate) \rightarrow CO_2 + H_2 \quad (28)$$

4. Sustainability

Syngas is acclaimed to be the energy of the future [7]. Hydrogen on the other hand has found application as a direct fuel in hydrogen internal combustion engine, an improvement from its initial application in fuel cells and energy storage. The sustainability of syngas as a fuel could be argued in accordance with its starting material, environmental footprint and economic implication. Interestingly, syngas can be produced from petroleum residua, coal, biomass and a number of opportunity fuels such as industrial and municipal wastes. The negative environmental impacts of these materials cannot be overstressed as much as the positive environmental impact of burning pure hydrogen coupled with growing technology of carbon capture, ultimately leading to the green technology of syngas as alternative fuel. However, the use of biomass and carbonaceous waste as feedstock would require tar removal [44], coupled with the fact that the reaction mechanism of the Fischer-Tropsch synthesis is still not well understood would negatively impact the economic implication of producing liquid fuel from syngas.

Alternatively, heavy crude oil and bitumen resources and technology for upgrading them is copious. Moreover, the literature is stacked with established reaction mechanisms for supplying hydrogen to aid hydrocracking via water gas shift reaction (or syngas). Although, they are finite non-renewable energy resources; their abundance makes them inexpensive, utilizing in-situ water gas shift reaction is economical and produce far less environmental footprint than supplying pure hydrogen, and their technology is comprehensible and accessible. In essence, syngas coupled with heavy crude oil and bitumen upgrading could provide near-term global energy independence, while facilitating the transition to a more sustainable syngas fuel.

5. Conclusion

Water gas shift reaction is an important step in the production of syngas. Interestingly, the sustainability of syngas span across the raw materials, process and application in fuel production as alternative to current and future energy demands. Worthy of note is its application in upgrading over 8 trillion barrels heavy crude oil and bitumen resources via water gas shift reaction to produce low molecular weight hydrocarbon separated into gases, liquefied petroleum gas (LPG), gasoline, naphtha, kerosene, diesel, gas oil, lubricating oil, greases and so on in a conventional refinery. For this purpose, hydrogen is produced in-situ from syngas with a suitable oxide catalysts and water from dehydrated or non-dehydrated crude. Upgrading via syngas gave lower viscosity and higher saturates. The process is also economically viable and environmentally friendly because of the absence of hydrogen gas in the feedstock and lower hydrogen sulfide in the gaseous product stream respectively. Although, heavy crude oil and bitumen are non-renewable fossil fuels, their abundant reserves and method of in-situ hydrogen generation during upgrading qualifies them as future alternative fuel and sustainable resource respectively.

Conflict of interest

The authors declare no conflict of interest.

Author details

Onoriode P. Avbenake[1,2]*, Faruk Ibrahim Yakasai[1,3] and Baba Y. Jibril[2]

1 Chemical and Petroleum Engineering Department, Bayero University, Kano, Nigeria

2 Chemical Engineering Department, Ahmadu Bello University, Zaria, Nigeria

3 University of Technology Malaysia, Johor, Malaysia

*Address all correspondence to: paulavbenake@gmail.com

References

[1] Richard P, Klaus D. Syngas use in internal combustion engines—A review. Advances in Research. 2017;**10**(1):1-8. DOI: 10.9734/AIR/2017/32896

[2] Loosdrecht J, Niemantsverdriet J. Synthesis gas to hydrogen, methanol, and synthetic fuels. In: Robert S, editor. Chemical Energy Storage. Germany: De Gruyter; 2012. DOI: 10.1515/9783110266320.443

[3] Guo K, Hailong L, Zhixin Y. In-situ heavy and extra-heavy oil recovery: A review. Fuel. 2016;**185**:886-902. DOI: 10.1016/j.fuel.2016.08.047

[4] Shokrlu Y, Babadagli T. Viscosity reduction of heavy oil/bitumen using micro- and nano-metal particles during aqueous and non-aqueous thermal applications. Journal of Petroleum Science and Engineering. 2014;**119**: 210-220. DOI: 10.1016/j.petrol.2014.05.012

[5] Faergestad I. Heavy oil. Oilfield Reviews. Paris, France: Schlumberger; 2016

[6] Ortiz-Moreno H, Ramírez J, Cuevas R, Marroquín G, Ancheyta J. Heavy oil upgrading at moderate pressure using dispersed catalysts: Effects of temperature, pressure and catalytic precursor. Fuel. 2012;**100**:186-192. DOI: 10.1016/j.fuel.2012.05.031

[7] Klasson K, Ackerson M, Clausen C, Gaddy L. Biological conversion of synthesis gas into fuels. International Journal of Hydrogen Energy. 1992; **17**(4):281-288. DOI: 10.1016/0360-3199(92)90003-F

[8] Nucara P, Sayma A. Effects of using hydrogen-rich syngas in industrial gas turbines while maintaining fuel flexibility on a multistage axial compressor design. In: Proceeding ASME Turbo Expo 2012: Turbine Technical Conference and Exposition. June 11–15. Copenhagen, Denmark: American Society of Mechanical Engineers (ASME); Paper number GT2012-69780; 2012. pp. 511-522. DOI: 10.1115/GT2012-69780

[9] Shekhawat D, Berry D, Spivey J. Introduction to fuel processing. In: Shekhawat D, Berry D, Spivey J, editors. Fuel Cells: Technologies for Fuel Processing. Oxford, UK: Elsevier; 2011. pp. 1-9. DOI: 10.1016/B978-0-444-53563-4.10001-X

[10] Chaudhari S, Dalai A, Bakhshi N. Production of hydrogen and/or syngas (H_2 + CO) via steam gasification of biomass-derived chars. Energy & Fuels. 2003;**17**:1062-1067. DOI: 10.1021/ef030017d

[11] Keyser M, Prinsloo F. Loading of cobalt on carbon nanofibers. Studies in Surface Science and Catalysis. 2007;**163**: 45-73. DOI: 10.1016/S0167-2991(07)80472-1

[12] Premkartikkumar S, Pradeepkumar A. Effect of using hydrogen mixed gases as a fuel in internal combustion engines—A review. International Journal of Innovative Research in Advanced Engineering. 2015;**2**(9):27-31

[13] Onoriode A, Rashid S, Baba J. Catalytic upgrading of heavy oil using $NiCo/\gamma\text{-}Al_2O_3$ catalyst: Effect of initial atmosphere and water-gas shift reaction. Fuel. 2019;**235**:736-743. DOI: 10.1016/j.fuel.2018.08.074

[14] Ramakanta S, Byung S, Sun J, Young-Pyo J, Chul W. A review of recent advances in catalytic hydrocracking of heavy residues. Journal of Industrial and Engineering Chemistry. 2015;**27**:12-24. DOI: 10.1016/j.jiec.2015.01.011

[15] Angeles M, Leyva C, Ancheyta J, Ramírez S. A review of experimental procedures for heavy oil hydrocracking with dispersed catalyst. Catalysis Today. 2014;**220-222**:274-294. DOI: 10.1016/j.cattod.2013.08.016

[16] Mohanty S, Kunzru D, Saraf D. Hydrocracking: A review. Fuel. 1990; **69**(12):1467-1473. DOI: 10.1016/0016-2361(90)90192-S

[17] Hossain M, Al-Saleh M, Shalabi M, Kimura T, Inui T. Pd–Rh promoted Co/HPS catalysts for heavy oil upgrading. Applied Catalysis A: General. 2004; **278**(1):65-71. DOI: 10.1016/j.apcata.2004.09.026

[18] Parkhomchuk E, Lysikov A, Okunev A, Parunin P, Semeikina V, Ayupov A, et al. Meso/macroporous CoMo alumina pellets for hydrotreating of heavy oil. Industrial and Engineering Chemistry Research. 2013;**52**:17117-17125. DOI: 10.1021/ie4022557

[19] Rana M, Sámano V, Ancheyta J, Diaz J. A review of recent advances on process technologies for upgrading of heavy oils and residua. Fuel. 2007; **86**(9):1216-1231. DOI: 10.1016/j.fuel.2006.08.004

[20] Mironenko O, Sosnin G, Eletskii P, Gulyaeva Y, Bulavchenko O, Stonkus O, et al. A study of the catalytic steam cracking of heavy crude oil in the presence of a dispersed molybdenum-containing catalyst. Petroleum Chemistry. 2017;**57**:618-629. DOI: 10.1134/S0965544117070088

[21] Omajali J, Hart A, Walker M, Wood J, Macaskie L. In-situ catalytic upgrading of heavy oil using dispersed bionanoparticles supported on gram-positive and gram-negative bacteria. Applied Catalysis B: Environmental. 2017;**203**:807-819. DOI: 10.1016/j.apcatb.2016.10.074

[22] Alemán-Vázquez L, Torres-Mancera P, Ancheyta J, Ramírez-Salgado J. On the use of hydrogen donors for partial upgrading of heavy petroleum. Energy & Fuels. 2016;**30**(11): 9050-9060. DOI: 10.1021/acs.energyfuels.6b01656

[23] Levin D, Chahine R. Challenges for renewable hydrogen production from biomass. International Journal of Hydrogen Energy. 2010;**35**(10): 4962-4969. DOI: 10.1016/j.ijhydene.2009.08.067

[24] Li J, Wang X, Tang X, Zhang M, Zheng X, Wang C, et al. Upgrading of heavy oil by thermal treatment in the presence of alkali-treated Fe/ZSM-5, glycerol, and biomass. Fuel Processing Technology. 2019;**188**:137-145. DOI: 10.1016/j.fuproc.2019.02.010

[25] Dejhosseini M, Aida T, Watanabe M, Takami S, Hojo D, Aoki N, et al. Catalytic cracking reaction of heavy oil in the presence of cerium oxide nanoparticles in supercritical water. Energy & Fuels. 2013;**27**:4624-4631. DOI: 10.1021/ef400855k

[26] Hata K, Wada K, Mitsudo T. Iron-catalyzed coprocessing of coals and vacuum residues using syngas-water as a hydrogen source. Energy & Fuels. 1998;**12**:1181-1190. DOI: 10.1021/ef980032w

[27] El harfi K, Mokhlisse A, Ben Chanaˆa M. Effect of water vapor on the pyrolysis of the Moroccan (Tarfaya) oil shale. Journal of Analytical and Applied Pyrolysis. 1999;**48**:65-76. DOI: 10.1016/S0165-2370(98)00108-9

[28] Zou R, Xu J, Kueffner S, Becker J, Li T, Guan X, et al. Spherical poly(vinyl imidazole) brushes loading nickel cations as nano catalysts for aquathermolysis of heavy crude oil. Energy Fuels. 2019;**33**:998-1006. DOI: 10.1021/acs.energyfuels.8b03964

[29] Chen M, Li C, Li G, Chen Y. Zhou C, In situ preparation of well-dispersed CuO nanocatalysts in heavy oil for

catalytic aquathermolysis. 2019;**16**: 439-446

[30] Siewe C, Ng F. Hydrodesulfurization of cold lake diesel fraction using dispersed catalysts: Influence of hydroprocessing medium and sources of H_2. Energy & Fuels. 1998;**12**:598-606. DOI: 10.1021/ef970188g

[31] Sitnov S, Mukhamatdinov I, Vakhin A, Ivanova A. Voronina E, Composition of aquathermolysis catalysts forming in situ from oil-soluble catalyst precursor mixtures. 2018;**169**:44-50

[32] Fumoto E, Sato S, Takanohashi T. Production of light oil by oxidative cracking of oil sand bitumen using iron oxide catalysts in a steam atmosphere. Energy & Fuels. 2011;**25**:524-527. DOI: 10.1021/ef101069m

[33] Chao K, Chen Y, Liu H, Zhang X, Li J. Laboratory experiments and field test of a difunctional catalyst for catalytic aquathermolysis of heavy oil. Energy & Fuels. 2012;**26**:1152-1159. DOI: 10.1021/ef2018385

[34] Rostrup-Nielsen J, Christiansen L. Concepts in syngas manufacture. In: Catalytic Science Series. Vol. 10. World Scientific. London, UK: Imperial college press; 2011. DOI: 10.1142/97818481 65687_0001

[35] Jeffrey W, Richard K. Downhole heavy crude oil hydroprocessing. Applied Catalysis A: General. 1996; **140**(1):1-16. DOI: 10.1016/0926-860X (96)00003-8

[36] Jeffrey W, Kessler R, Sawicki R, Belgrave J, Laureshen C, Mehta S, et al. Down-hole catalytic upgrading of heavy crude oil. Energy & Fuels. 1996;**10**(4): 883-889. DOI: 10.1021/ef9501814

[37] Sato T, Tomita T, Trung P, Itoh N, Sato S, Takanohashi T. Upgrading of bitumen in the presence of hydrogen and carbon dioxide in supercritical water. Energy & Fuels. 2013;**27**:646-653. DOI: 10.1021/ef301444j

[38] Fumoto E, Tago T, Masuda T. Production of lighter fuels by cracking petroleum residual oils with steam over zirconia-supporting iron oxide catalysts. Energy & Fuels. 2006;**20**(1):1-6. DOI: 10.1021/ef050105t

[39] Ajumobi O, Muraza O, Kondoh H, Hasegawa N, Nakasaka Y, Yoshikawa T, et al. Upgrading oil sand bitumen under superheated steam over ceria-based nanocomposite. Catalysts. 2018;**218**:1-9. DOI: 10.1016/j.apenergy.2018.02.161

[40] Hosseinpour M, Fatemi S, Ahmadi J, Morimoto M, Akizuki M, Oshima Y, et al. The synergistic effect between supercritical water and redox properties of iron oxide nanoparticles during in-situ catalytic upgrading of heavy oil with formic acid. Isotopic study. Applied Catalysis B: Environmental. 2018;**230**:91-101. DOI: 10.1016/j.apcatb.2018.02.030

[41] Nguyen N, Blum L. Syngas and synfuels from H_2O and CO_2: Current status. Chemie Ingenieur Technik. 2015; **87**(00):1-23. DOI: 10.1002/cite.201400090

[42] Cinti G, Baldinelli A, Di Michele A, Desideri U. Integration of solid oxide electrolyzer and Fischer-Tropsch: A sustainable pathway for synthetic fuel. Applied Energy. 2016;**162**:308-320

[43] König DH, Freiberg M, Dietrich R-U, Wörner A. Techno-economic study of the storage of fluctuating renewable energy in liquid hydrocarbons. Fuel. 2015;**159**:289-297

[44] Speight J. Production of syngas, synfuel, bio-oils, and biogas from coal, biomass, and opportunity fuels. In: Oakey J, editor. Fuel Flexible Energy Generation. Amsterdam, The Netherlands: Woodhead Publishing, Elsevier; 2016. pp. 145-174. DOI: 10.1016/B978-1-78242-378-2.00006-7

[45] Schemme S, Samsun R, Peters R, Stolten D. Power-to-fuel as a key to sustainable transport systems—An analysis of diesel fuels produced from CO2 and renewable electricity. 2017; **205**:198-221. DOI: 10.1016/j.fuel.2017.05.061

[46] Stiles A. Methanol, past, present, and speculation on the future. AICHE Journal. 1977;**23**(3):362-375. DOI: 10.1002/aic.690230321

[47] Stiles A, Chen F, Harrison J, Hu X, Storm D, Yang H. Catalytic conversion of synthesis gas to methanol and other oxygenated products. Industrial and Engineering Chemistry Research. 1991; **30**(5):811-821. DOI: 10.1021/ie00053a002

[48] Reubroycharoen P, Yamagami T, Vitidsant T, Yoneyama Y, Ito M, Tsubaki N. Continuous low-temperature methanol synthesis from syngas using alcohol promoters. 2003;**17**:817-821. DOI: 10.1021/ef020240v

[49] Ashik U, Daud W, Abbas H. Production of greenhouse gas free hydrogen by thermocatalytic decomposition of methane—A review. Renewable and Sustainable Energy Reviews. 2015;**44**:221-256. DOI: 10.1016/j.rser.2014.12.025

[50] Apak S, Atay E, Tuncer G. Renewable hydrogen energy regulations, codes and standards: Challenges faced by an EU candidate country. International Journal of Hydrogen Energy. 2012;**37**(7): 5481-5497. DOI: 10.1016/j.ijhydene.2012.01.005

[51] Dahiya P, Chand A. Utilisation of hydrogen for domestic, commercial and industrial applications. In: Dahiya P, editor. Progress in Hydrogen Energy. Netherlands: Springer; 1987. pp. 179-194. DOI: 10.1007/978-94-009-3809-0_12

[52] Laan G, Beenackers A. Kinetics and selectivity of the Fischer–Tropsch synthesis: A literature review. Catalysis Reviews. 1999;**41**:255-318. DOI: 10.1081/CR-100101170

[53] Hewu W, Haiyan H, Xue D, Minggao O. Hydrogen and fuel-cell vehicle technology. In: CAERC, editor. Sustainable Automotive Energy System in China. Berlin Heidelberg: Springer; 2013. pp. 301-333. DOI: 10.1007/978-3-642-36847-9_10

[54] Baharudin L, Watson M. Hydrogen applications and research activities in its production routes through catalytic hydrocarbon conversion. Reviews in Chemical Engineering. 2017;**34**:43-72. DOI: 10.1515/revce-2016-0040

[55] Arcelus-Arrillaga P, Pinilla J, Hellgardt K, Millan M. Application of water in hydrothermal conditions for upgrading heavy oils: A review. Energy & Fuels. 2017;**31**:4571-4587. DOI: 10.1021/acs.energyfuels.7b00291

[56] Sato T, Kurosawa S, Smith R, Adschiri T, Arai K. Water gas shift reaction kinetics under noncatalytic conditions in supercritical water. Journal of Supercritical Fluids. 2004;**29** (1–2):113-119. DOI: 10.1016/S0896-8446(03)00049-4

[57] Ruilin R, Huachao L, Yu C, Jian L, YanLing C. Improving the aquathermolysis efficiency of aromatics in extraheavy oil by introducing hydrogen-donor ligands to catalysts. Energy & Fuels. 2015;**29**(12):7793-7799. DOI: 10.1021/acs.energyfuels.5b01256

[58] Araki K, Fujiwara H, Sugimoto K, Oshima Y, Koda S. Kinetics of water-gas shift reaction in supercritical water. Journal of Chemical Engineering of Japan. 2004;**37**(3):443-448. DOI: 10.1252/jcej.37.443

[59] Rodriguez J, Ma S, Liu P, Hrbek J, Evans J, Pérez M. Activity of CeO_x and

TiO_x nanoparticles grown on Au (111) in the water-gas shift reaction. Science. 2007;**318**:1757-1760. DOI: 10.1126/science.1150038

[60] Hajdo L, Hallam R, Vorndran L. Hydrogen generation during in situ combustion. In: Proceedings of the SPE California Regional Meeting. 27-29 March. Bakersfield, California: Society of Petroleum Engineers (SPE); SPE Paper 13661; 1985. DOI: 10.2118/13661-MS

[61] Adschiri T, Shibata R, Sato T, Watanabe M, Arai K. Catalytic hydrodesulfurization of dibenzothiophene through partial oxidation and a water-gas shift reaction in supercritical water. Industrial and Engineering Chemistry Research. 1998; **37**:2634-2638. DOI: 10.1021/ie970751i

[62] Kapadia P, Kallos M, Gates I. Potential for hydrogen generation from in situ combustion of Athabasca bitumen. Fuel. 2011;**90**:2254-2265. DOI: 10.1016/j.fuel.2011.02.038

[63] Thomas F, Moore R, Bennion D. Kinetics parameters for the high temperature oxidation of in-situ combustion coke. Journal of Canadian Petroleum Technology. 1985;**24**(6): 60-67. DOI: 10.2118/85-06-05

[64] Hart A, Leeke G, Greaves M, Wood J. Downhole heavy crude oil upgrading using CAPRI: Effect of steam upon upgrading and coke formation. Energy & Fuels. 2014;**28**:1811-1819. DOI: 10.1021/ef402300k

[65] Maity S, Ancheyta J, Marroquin G. Catalytic aquathermolysis used for viscosity reduction of heavy crude oils: A review. Energy & Fuels. 2010;**24**: 2809-2816. DOI: 10.1021/ef100230k

[66] Smith B, Loganathan M, Shantha M. A review of the water gas shift reaction kinetics. International Journal of Chemical Reactor Engineering. 2010;**8**: 1-32. DOI: 10.2202/1542-6580.2238

Section 2

Syngas Fuel Production

Chapter 4

Syngas Fuel Production from Carbonaceous Feedstocks Using Hybrid Porous Media

Mario Toledo Torres and Nicolás Ripoll Kameid

Abstract

During the last years, hybrid porous media reactors have been developed aiming to partially oxidize solid and gaseous fuels to produce reducing gases. The gases produced are mainly composed of hydrogen (H_2) and carbon monoxide, among other products of gasification. This hybrid process combines inert porous media (IPM) combustion and gasification of solid fuels by replacing a fraction of the inert solid volume with a solid fuel. The gaseous mixture is produced from carbon-rich reactants exposed to the high temperatures of filtration combustion. Experimental results from different solid fuels (coal, biomass, and others) and gaseous fuels (natural gas (NG), propane, and others) are presented, with detailed analysis of high temperatures (between 900 and 1800 K), velocities, and product gas composition of the combustion waves, which is able to produce $[H_2]/[CO]$ ratios from 0.2 to 10.

Keywords: hydrogen production, solid and gaseous fuels, porous media, hybrid filtration combustion

1. Introduction

Hydrogen (H_2) and syngas production technology development has been concentrating most of current efforts toward more efficient and responsible use of fossil carbonaceous feedstocks [1, 2]. Moreover, these technologies can utilize energy more efficiently, supply ultraclean fuels, eliminate pollutant emissions at end-use systems, and significantly reduce greenhouse gases emissions, particularly carbon dioxide (CO_2) [2]. H_2 and syngas are currently mostly produced by steam reforming, partial oxidation, and autothermic reforming which is also known as oxidative steam reforming [2]. For example, H_2 is mainly produced by steam methane reforming (SMR), a process that inherently releases huge amounts of greenhouse gases. The primary energy sources to produce hydrogen are hydrocarbon feedstocks (methane, oil, and coal) with 96% of the supply, while the rest (4%) is attained through water electrolysis [3]. However, in past years, it has been challenging to properly forecast the availability of hydrocarbon feedstocks, which in turn adds to its uncertainty as a main feedstock in the H_2 production chain. Therefore, the development of novel techniques aimed to diversify H_2 and syngas production presents itself as highly necessary, where the gasification of biomass, for example, poses as a promising effort to significantly compete against fossil feedstocks [4, 5], with a carbon-neutral alternative.

Several applications, processes, and configurations have been developed to thermochemically transform solid fuels into a gaseous fuel through a process called gasification. This process consists on the transformation of solid substances that contain carbon such as biomass, coal, or waste into a combustible gaseous product in the presence of air, water steam ($H_2O_{(g)}$), oxygen (O_2), CO_2, or a mixture of these gasifying agents. The conversion of these substances occurs at high temperatures (~800°C) and moderated pressures (from atmospheric pressure of up to 70 barg) [6–8]. The conventional gasifiers are classified based on the type of bed and direction of the gas flow [9]; the description of their functioning principles and main features can be extensively found in technical literature [10]. In particular, biomass gasification differs from coal gasification, mainly because biomass is a carbon-neutral and sustainable energy source and because biomass is more reactive and features a higher volatile content than coal, which results in a lower gasification temperature. This reduces heat loss, undesired emissions, and material problems associated with high temperatures. Biomass also has a low sulfur content, which results in less SO_X emissions, but due to its high alkali contents, like sodium and potassium, slagging and fouling are common problems in biomass gasification equipment [11]. There are several studies regarding solid fuel gasification, such as the results reported by [12–14]. On the other hand, disadvantages of catalytic gasification include increased material costs for the catalyst (often rare metals), as well as diminishing catalyst performance over time. The relative difficulty in reclaiming and recycling the catalyst can also be a disadvantage [8].

In general terms, gasification as a process still requires further optimizations to enhance its energy efficiency by overcoming the main aforementioned challenges, such as tar production and moisture content of the biomass. Although new technologies have been developed as effective ways to utilize even toxic and wet biomass for power generation [15] and conventional techniques have been proven to provide a feasible option to reform solid fuels, there are still limitations on the characteristics of the fuel that restrict the use to certain feedstocks. Fixed bed gasifiers may work with solid fuels containing up to 50% of humidity, while fluidized bed gasifiers can work with solid fuels with up to 60% of humidity in the most advanced developments [16]. The products obtained in the different configurations of gasifier devices are mainly composed of H_2, CO, CO_2, H_2O, N_2, heavier hydrocarbons (C_2–C_6), ashes, tar, oils, and small solid carbon particles, among others. Finally, the main disadvantages of conventional autothermic gasification technology are related to the production of undesired species such as particulate matter, tar, and char. The emissions of these species are highly associated to the operational parameters of the process such as temperature, pressure, time and heating speed, solid fuel particle size, and residence time [17]. For these reasons, researchers have studied the technology of inert porous media (IPM) combustion detecting many important advantages, such as low pollutant emissions, high thermal stability, increased reaction temperature due to its internal heat recirculation, and extended flammability limits, among others [18–21].

The main objective of this chapter is to present the use of IPM technology for achieving high-temperature gasification of solid fuel in a hybrid porous media reactor.

2. Hybrid filtration combustion for solid fuels

IPM is a thermochemical process proven to be a feasible option to address current global requirements for cleaner energy sources and processes [22]. This technology is known to be able to produce H_2 from several feedstocks and allows the direct use of liquid and gaseous fuels that interact with an inert solid matrix.

The common approaches to use the technology are the stationary and transient configurations.

A relation between IPM combustion and solid fuel gasification converges into hybrid filtration combustion (HFC), a process that combines the properties of the aforesaid processes by replacing a fraction of the inert solid's volume with a solid fuel. In this case, a reaction wave is produced by a flow that can contain hot air, H_2O(g), or a gaseous fuel-air mixture that propagates along the reactor reforming the solid fuel inside within a wide-power-range, high-efficiency, high energy concentration per unit of volume and stable combustion over a wide range of equivalence ratios [23]. Several experimental studies on HFC for syngas and H_2 production have been conducted [24–35], showing that the technology presents a strong and feasible option for syngas production from gaseous and solid fuels in a batch configuration.

In [24] three types of algae were analyzed, showing that an increase of volume algae fraction in the hybrid bed and an increase of moisture content in the algae used increased both combustion temperature and hydrogen yields. Different gasifying agents were used on experiments with biomass pellets and alumina spheres using equal volumetric fractions [25]. While operating with natural gas (NG), the combustion wave temperature increased only using insignis pine, whereas the usage of cereal plantation residuals enhanced the syngas production. Using steam, the combustion wave temperature presented a slight decrease as the steam presence increased. In the case of natural gas in a porous medium composed of coal and alumina particles [26], the flame temperature decreased with an increase of coal fraction, and hydrogen and carbon monoxide were dominant partial oxidation products. Further experimental studies [29, 31] consistently reported that hydrogen and carbon monoxide are dominant partial oxidation products for atmospheric hybrid combustion waves.

Industrial applications of HFC have been successfully implemented in Northern-European countries, such as Finland and Russia, where two reactors capable of processing up to 15,000 ton/yr of municipal solid wastes (MSW) were engineered by the IPCP-RAS (Russia) and developed by Europrofile Ltd. in Lappeenranta (Finland) and Moscow (Russia) [36, 37].

3. Experimental results by hybrid filtration combustion

Figure 1 shows the experimental setup generally used in hybrid porous media reactors. The filtration combustion system consists of a tube, usually made of quartz, filled with uniformly mixed aleatory ceramic spheres and solid fuel particles. To compensate for the different thermal expansion rates of the packed bed and tube, the inside diameter of the tube is covered with an insulation blanket (ceramic fiber). Heat losses due to conduction through the tube wall are minimized with an additional insulation layer covering the outside of the tube. Air, fuel, and/or steam, metered using mass flow controllers, are premixed before entering the reactor and introduced into the reactor from its bottom. The upstream or downstream propagating combustion wave is ignited using a lighter at the reactor exit or reactor bottom. System diagnostics are required to assess the temperature profile in the reactor and the chemical composition of the output gases. The axial temperature distribution of the reactor is measured by thermocouples. These thermocouples are housed in a multi-bored ceramic shell. A data acquisition system is used to read and record the temperatures. The digital conversion of the resultant analog signals is performed with a data acquisition board. Finally, the chemical composition of the flue gases is measured using a gas chromatograph.

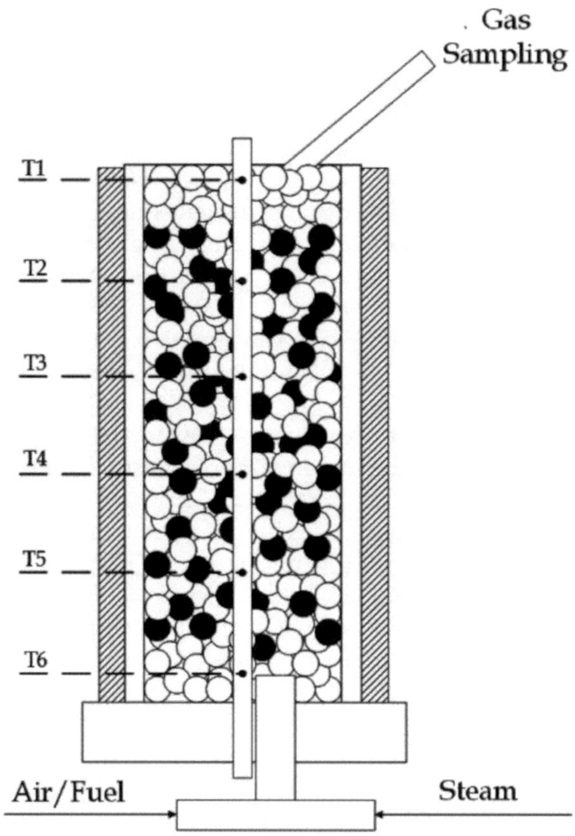

Figure 1.
Typical experimental setup.

In **Figure 2**, the experimental results for measured combustion temperatures and propagation rates, and hydrogen and carbon monoxide concentrations, in the form of the ratio $[H_2]/[CO]$, are displayed for a range of equivalence ratio (φ) of $0.3 < \varphi < 2.6$. This particular graph displays profiles belonging to three different gaseous and solid fuel mixtures in hybrid process as natural gas and wood, propane and polyethylene, and butane and wood. All experiments were conducted on a hybrid bed composed of 50% of solid fuel and 50% inert alumina spheres.

In **Figure 2A**, the combustion temperatures range is from 1,000 to 1,300 K for lean mixtures ($\varphi < 1.0$), from 1,100 to 1,200 K in rich mixtures ($1.0 < \varphi < 1.65$), and from 1,200 to 1,500 K in ultrarich mixtures ($\varphi > 1.3$). The high stable combustion temperatures evidenced in the range of $1,000 < \varphi < 1,500$ show that hybrid process feasibility is almost independent of the gaseous and solid fuels for the equivalence ratio experimented. The high overall combustion temperature represents that the reactors provide suitable conditions for the combustion chemistry to convert simultaneously gaseous and solid fuels into synthesis gas. This figure reveals some interesting characteristics of transient filtration combustion waves depicting the regions of wave propagating counter or concurrent to the unburned gas. At $\varphi < 0.4$ upstream superadiabatic wave is observed. If more gaseous fuel is added in the mixtures, stable underadiabatic waves are developed. Underadiabatic waves are established from approximately $\varphi = 0.4$ to $\varphi = 2.4$. Further addition of gaseous fuels develops upstream superadiabatic waves at $\varphi > 2.4$.

In **Figure 2B**, the products of combustion comparatively show $[H_2]/[CO]$ ratio. It was observed that lean mixtures with higher oxygen content in the oxidizer stream

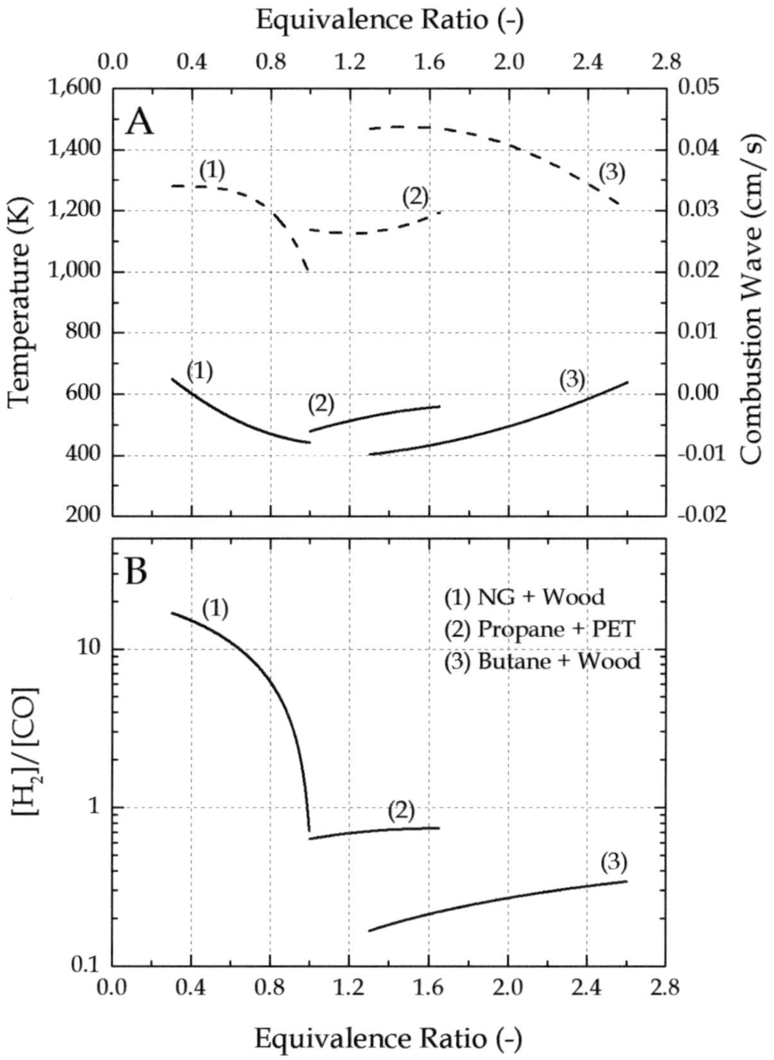

Figure 2.
Effect of varying the equivalence ratio on (A) maximum recorded temperatures inside the reactor (dashed lines) and combustion wave propagation rates (solid lines) and (B) [H_2]/[CO] ratios (solid lines), for three different experimental sets: (1) ultralean natural gas and air mixtures with wood pellets [30], (2) rich propane-air mixtures with polyethylene [31], and (3) rich and ultrarich butane-air mixtures with wood pellets [29].

generate higher amounts of [H_2]/[CO] ratios. The overall region of high ratio is between $0.3 < \varphi < 1.0$. In rich and ultrarich mixtures, the [H_2]/[CO] ratios are less than 1. Of utmost importance in this chapter is the adequate ratio in the combustion products, since it will determine the usefulness of the obtained syngas in the desired applied process. From lean to rich regimes, the products of combustion are controlled by temperature and residence time. The process can be characterized as gaseous and solid fuel reforming rather than combustion such that a stable thermal process exists with a concentration of oxygen. This type of hybrid system is possible by the unconstrained movement of the combustion zone to recuperate its energy from the porous matrix.

An important operational parameter of the hybrid filtration combustion reactor is the mass fraction of solid fuels in the inert porous matrix. **Figure 3** shows experimental results for temperatures, combustion waves, and [H_2]/[CO] ratios. It was found that the combustion temperature remains higher (900–1,700 K) in all the range

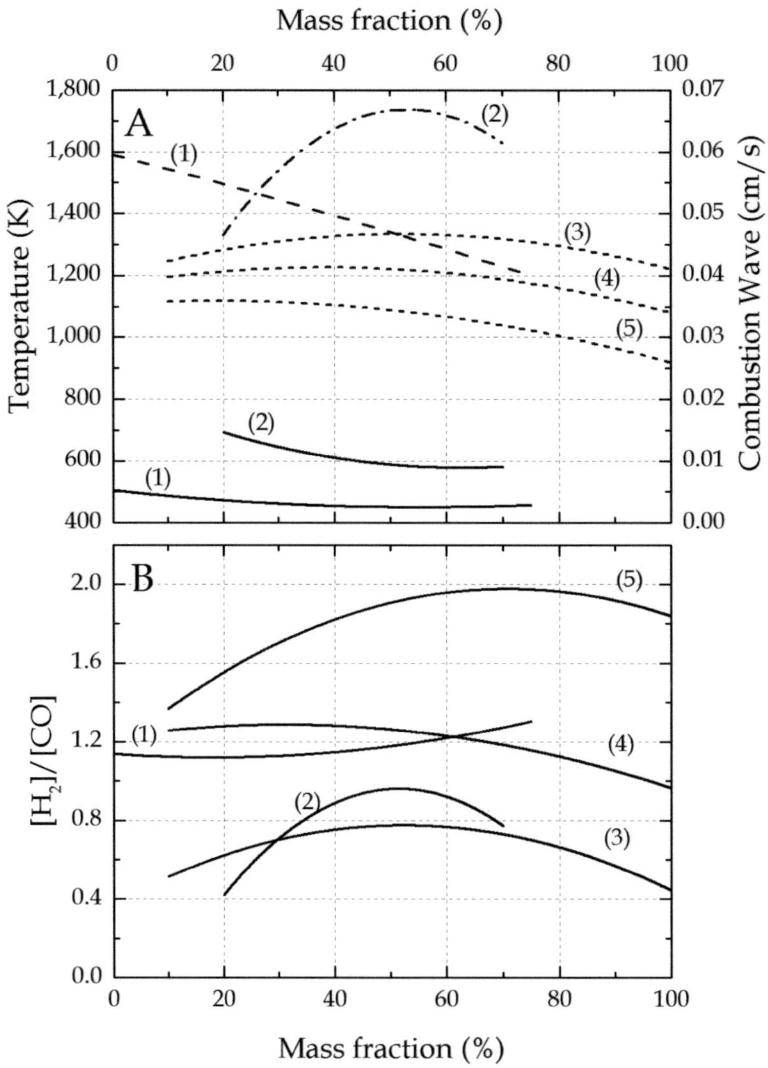

Figure 3.
Effect of varying the mass fraction of different carbonaceous fuels in the porous matrix on the (A) peak recorded temperatures (dashed, dotted, and dash-dotted lines) and combustion wave propagation rate (solid lines) and (B) $[H_2]/[CO]$ ratios for three different experimental sets: (1) ultrarich natural gas and air mixture with bituminous coal [26], (2) steam-air mixtures with bituminous coal [23], and (3) charcoal with different fractions of $[H_2O]/[O_2]$ [38]: (1) 1.0; (2) 2.5; and (3) 4.5.

of mass fraction presented. In high mass fraction (> 60%), temperature decreases because heat is necessary to convert solid fuels into syngas. The combustion waves for (1) and (2) were found to be totally superadiabatic over the whole range of mass fraction tested. With different gaseous and solid fuels, the $[H_2]/[CO]$ ratios formed are between 0.4 and 2.0, in all the range of mass fraction reported (**Figure 3B**).

4. Conclusion

In this chapter, experimental results were presented for hybrid filtration combustion of different gaseous and solid fuels. Results are focused in combustion temperatures, waves, and $[H_2]/[CO]$ ratios with varying equivalence ratio and mass fraction of solid fuels in the inert porous matrix.

Depending on operational parameters, the range of high combustion temperature is between 900 and 1,800 K. Considering that the $[H_2]/[CO]$ ratios are from 0.2 to 10, the applications for the hybrid reactor will depend on the use of this ratio in the next applied process.

Acknowledgements

The authors wish to acknowledge the financial support by the FONDECYT 1190654 and FONDAP 15110019/SERC-Chile.

Author details

Mario Toledo Torres* and Nicolás Ripoll Kameid
Department of Mechanical Engineering, Universidad Técnica Federico Santa María, Valparaíso, Chile

*Address all correspondence to: mario.toledo@usm.cl

References

[1] Jamal Y, Wyszynski ML. On-board generation of hydrogen-rich gaseous fuels-a review. International Journal of Hydrogen Energy. 1994;**19**:557-572. DOI: 10.1016/0360-3199(94)90213-5

[2] Liu K, Song C, Subramani V. Hydrogen and Syngas Production and Purification Technologies. 1st ed. Hoboken, New Jersey: John Wiley & Sons, Inc; 2010

[3] Ewan BCR, Allen RWK. A figure of merit assessment of the routes to hydrogen. International Journal of Hydrogen Energy. 2005;**30**:809-819. DOI: 10.1016/j.ijhydene.2005.02.003

[4] Bhat A, Ram Bheemarasetti JV, Rajeswara RT. Kinetics of rice husk char gasification. Energy Conversion and Management. 2001;**42**:2061-2069. DOI: 10.1016/S0196-8904(00)00173-4

[5] Chen X, Honda K, Zhang ZG. A comprehensive comparison of CH4-CO2 reforming activities of NiO/Al2O3 catalysts under fixed- and fluidized-bed operations. Applied Catalysis A: General. 2005;**288**:86-97. DOI: 10.1016/j.apcata.2005.04.037

[6] El-Emam RS, Dincer I, Naterer GF. Energy and exergy analyses of an integrated SOFC and coal gasification system. International Journal of Hydrogen Energy. 2012;**37**:1689-1697. DOI: 10.1016/j.ijhydene.2011.09.139

[7] Ruiz JA, Juárez MC, Morales MP, Muñoz P, Mendívil MA. Biomass gasification for electricity generation: Review of current technology barriers. Renewable and Sustainable Energy Reviews. 2013;**18**:174-183. DOI: 10.1016/j.rser.2012.10.021

[8] Elsevier LtdLuque R, Speight JG. Part three: Applications. In: Gasification for Synthetic Fuel Production. 2015. pp. 201-320. DOI: 10.1016/C2013-0-16368-4

[9] Kumar A, Jones DD, Hanna MA. Thermochemical biomass gasification: A review of the current status of the technology. 2009:556-581. DOI: 10.3390/en20300556

[10] Higman C, van der Burgt M. Gasification. Elsevier; 2003. DOI: 10.1016/B978-075067707-3/50001-2

[11] Shah YT. Synthesis gas by thermal gasification. In: Chemical Energy from Natural and Synthetic Gas. 1st ed. Boca Raton: CRC Press; 2017. pp. 113-209

[12] Peng WX, Wang LS, Mirzaee M, Ahmadi H, Esfahani MJ, Fremaux S. Hydrogen and syngas production by catalytic biomass gasification. Energy Conversion and Management. 2017;**135**:270-273. DOI: 10.1016/j.enconman.2016.12.056

[13] Asadullah M, Miyazawa T, Ito SI, Kunimori K, Tomishige K. Demonstration of real biomass gasification drastically promoted by effective catalyst. Applied Catalysis A: General. 2003;**246**:103-116. DOI: 10.1016/S0926-860X(03)00047-4

[14] Chen Z, Dun Q, Shi Y, Lai D, Zhou Y, Gao S, et al. High quality syngas production from catalytic coal gasification using disposable Ca(OH)2 catalyst. Chemical Engineering Journal. 2017;**316**:842-849. DOI: 10.1016/j.cej.2017.02.025

[15] Farzad S, Mandegari MA, Görgens JF. A critical review on biomass gasification, co-gasification, and their environmental assessments. Biofuel Research Journal. 2016;**3**:483-495. DOI: 10.18331/BRJ2016.3.4.3

[16] Quaak P, Knoef H, Stassen H. Energy from Biomass: A

Review of Combustion and Gasification Technologies. 1999. DOI: 10.1596/0-8213-4335-1

[17] Mondal P, Dang GS, Garg MO. Syngas production through gasification and cleanup for downstream applications - Recent developments. Fuel Processing Technology. 2011;**92**:1395-1410. DOI: 10.1016/j.fuproc.2011.03.021

[18] Carvalho T, Costa M, Casaca C, Catapan RC, Oliveira AAM. Destruction of the tar present in syngas by combustion in porous media. Energy and Fuels. 2015;**29**:1130-1136. DOI: 10.1021/ef501807p

[19] Kamal MM, Mohamad AA. Combustion in porous media. Proceedings of the Institution of Mechanical Engineers, Part A: Journal of Power and Energy. 2006;**220**:487-508. DOI: 10.1243/09576509JPE169

[20] Mujeebu MA, Abdullah MZ, Bakar MZA, Mohamad AA, Abdullah MK. Applications of porous media combustion technology - A review. Applied Energy. 2009;**86**:1365-1375. DOI: 10.1016/j.apenergy.2009.01.017

[21] Wood S, Harris AT. Porous burners for lean-burn applications. Progress in Energy and Combustion Science. 2008;**34**:667-684. DOI: 10.1016/j.pecs.2008.04.003

[22] Bingue JP, Saveliev AV, Fridman AA, Kennedy LA. Hydrogen production in ultra-rich filtration combustion of methane and hydrogen sulfide. International Journal of Hydrogen Energy. 2002;**27**:643-649. DOI: 10.1016/S0360-3199(01)00174-4

[23] Toledo M, Araus SK, Vasconcelo AD. Syngas production from coal in presence of steam using filtration combustion. International Journal of Hydrogen Energy. 2015;**40**:6340-6345. DOI: 10.1016/j.ijhydene.2015.03.022

[24] Ripoll N, Silvestre C, Paredes E, Toledo M. Hydrogen production from algae biomass in rich natural gas-air filtration combustion. International Journal of Hydrogen Energy. 2017;**42**:1-10. DOI: 10.1016/j.ijhydene.2016.03.082

[25] Caro S, Torres D, Toledo M. Syngas production from residual biomass of forestry and cereal plantations using hybrid filtration combustion. International Journal of Hydrogen Energy. 2015;**40**:2568-2577. DOI: 10.1016/j.ijhydene.2014.12.102

[26] Toledo M, Utria KS, González FA, Zuñiga JP, Saveliev AV. Hybrid filtration combustion of natural gas and coal. International Journal of Hydrogen Energy. 2012;**37**:6942-6948. DOI: 10.1016/j.ijhydene.2012.01.061

[27] Salganskaya MV, Glazov SV, Salgansky EA, Kislov VM, Zholudev AF, Manelis GB. Filtration combustion of humid fuels. Russian Journal of Physical Chemistry B. 2008;**2**:71-76. DOI: 10.1134/S1990793108010119

[28] Kislov VM, Glazov SV, Salgansky EA, Kolesnikova YY, Salganskaya MV. Coal gasification by a mixture of air and carbon dioxide in the filtration combustion mode. Combustion, Explosion, and Shock Waves. 2016;**52**:320-325. DOI: 10.1134/S0010508216030102

[29] Toledo M, Vergara E, Saveliev AV. Syngas production in hybrid filtration combustion. International Journal of Hydrogen Energy. 2011;**36**:3907-3912. DOI: 10.1016/j.ijhydene.2010.11.060

[30] Araus K, Reyes F, Toledo M. Syngas production from wood pellet using filtration combustion of lean natural gas-air mixtures. International Journal

of Hydrogen Energy. 2014;**39**:7819-7825. DOI: 10.1016/j.ijhydene.2014.03.140

[31] Gentillon P, Toledo M. Hydrogen and syngas production from propane and polyethylene. International Journal of Hydrogen Energy. 2013;**38**:9223-9228. DOI: 10.1016/j.ijhydene.2013.05.058

[32] Toledo M, Rosales C. Hybrid filtration combustion. Hydrogen Energy - Challenges and Perspectives. 2012:201-222. DOI: 10.5772/50353

[33] Salganskii EA, Fursov VP, Glazov SV, Salganskaya MV, Manelis GB. Model of vapor-air gasification of a solid fuel in a filtration mode. Combustion, Explosion, and Shock Waves. 2006;**42**:55-62. DOI: 10.1007/s10573-006-0007-9

[34] Amelin II, Salgansky EA, Volkova NN, Zholudev AF, Alekseev AP, Polianczyk EV, et al. Parametric domain of the stationary filtration combustion wave in the charge with a low carbon content. Russian Chemical Bulletin. 2011;**60**:1150-1157. DOI: 10.1007/s11172-011-0180-1

[35] Salgansky EA, Polianchik EV, Manelis GB. Modeling filtration combustion of a pyrolyzing solid fuel. Combustion, Explosion, and Shock Waves. 2013;**49**:38-52. DOI: 10.1134/S001050821301005X

[36] Manelis GB, Glazov SV, Lempert DB, Salgansky EA. Filtration combustion of solid fuel in countercurrent reactors. Russian Chemical Bulletin. 2011;**60**:1301-1317. DOI: 10.1007/s11172-011-0198-4

[37] Kislov VM, Glazov SV, Chervonnaya NA, Patronova LI, Salganskaya MV, Manelis GB. Biomass gasification under combustion conditions with superadiabatic heating. Solid Fuel Chemistry. 2008;**42**:135-139. DOI: 10.3103/S0361521908030038

[38] Salgansky EA, Kislov VM, Glazov SV, Zholudev AF, Manelis GB. Filtration combustion of a carbon-inert material system in the regime with superadiabatic heating. Combustion, Explosion, and Shock Waves. 2008;**44**:273-280. DOI: 10.1007/s10573-008-0035-8

Chapter 5

Discussion on the Feasibility of Industrial Fuel Gas Prepared by Lurgi Gasification Which Used Anthracite as Raw Material

Zhang Min and Zhang Haifeng

Abstract

The long, stable operation of Jinshi and Tianqing programs proved the feasibility of Lurgi gasification by using Jincheng anthracite as raw material. Compared to other types of coal, Lurgi gasification by using Jincheng anthracite as raw material showed advantages including environmental protection, water consumption, effluent disposal, etc. By analyzing the characteristics of Lurgi gasification by using Jincheng anthracite as raw material, we discussed the choice of technical route and confirmed that the route of industrial gas, coal chemical products, and natural gas could exploit the advantage to the full. Finally, techno-economic evaluation of fuel gas prepared by Lurgi gasification with anthracite as raw material was discussed, and the result showed that industrial fuel gas produced by Lurgi gasification with anthracite as raw material could become an alternative to natural gas in industrial fuel gas field.

Keywords: fuel gas, Lurgi gasification, anthracite, techno-economic evaluation

1. Introduction

As a large corporate group, Jincheng Anthracite Coal Mining Group Corporation Limited (JAMG) mainly deals with the exploitation and sale of anthracite. In the past, when new coal gasification technology had not been industrialized, United Gas Improvement Company gasification (UGI) technology was the major gasification technology in coal chemical industry. Since the beginning of twenty-first century, many types of gasification technology designed for other types of coal has gradually industrialized and matured, such as HT-L gasification technology [1], SE pulverized coal gasification technology [2], GSP gasification technology [3], shell gasification technology [4], opposed multi-burner coal water slurry gasification technology [5], and Texaco coal-water slurry gasification technology [6]. The monopolistic status of anthracite as raw material in chemical industry has been broken. Expanding the application of anthracite has become the key issue for the sustainable development of JAMG.

In China, industrial fuel gas, which was produced by two-stage coal gasification technology and fluidized-bed coal gasification technology, is mainly applied in glass industry, ceramics industry, nonferrous metal industry, etc. However, with the development of society, the mode of production is no longer suitable eager for transformation and sustainable development. Production mode with characteristics of environment-friendly, supply-stable, and low cost is the development trend of industrial fuel gas [7, 8].

Natural gas, which is recognized as the cleanest fossil energy resource, has been vigorously promoted by the government in industrial fuel gas field. But as the characteristics of resource endowment in China are "rich coal, lack of oil, and less gas," the supply of natural gas in industrial fuel gas field is far from enough. How to acquire other types of industrial fuel gas which can meet with the development trend has become the key issue for the sustainable development of glass industry, ceramics industry, nonferrous metal industry, etc. [7–9].

In China, the resource endowment is rich in coal and lack of oil and gas. It is a good way to make full use of abundant coal resources to produce industrial fuel gas. According to the movement state of coal and gasifier in the gasifier, coal gasification technology has three kinds: fixed bed gasification technology, fluidized bed gasification technology, and entrained bed gasification technology [10, 11].

Fixed bed gasification technology mainly includes intermittent atmospheric fixed bed (UGI), Lurgi, BGL, etc. Intermittent atmospheric fixed bed is being phased out worldwide because of its high pollution and energy consumption. Lurgi furnace is made of crushed (lump) coal as raw material. The content of effective gas ($CO + H_2$) and methane is about 65 and 8–10%, respectively. It is suitable for producing industrial fuel gas. Compared with other kinds of coal, the tar, phenol, and ammonia produced from anthracite are less difficult to treat, and the amount of waste water is also less [10–12].

The representative types of fluidized bed are high temperature Winkler gasification technology and ash fusion gasification technology of Shanxi Coal Chemical Institute. Because the technology has many problems, such as low effective component, high impurity (high carbon content, difficult separation), high activity of coal, and high ash melting point, it is seldom used in China [10, 11].

At present, the entrained flow bed is the main choice for large-scale production. The representative furnace type is shell, space furnace (HT-L), Texaco, etc. Its characteristics are high reaction temperature (1500–2000°C) and high conversion efficiency. Its effective gas composition ($CO + H_2$) is as high as 85–92%, while CH_4 composition is very low or almost no, and its calorific value is relatively low, so it is not suitable for the production of industrial fuel gas [10, 11, 13].

So, the application as an industrial fuel gas by taking Lurgi gasification, which used anthracite as raw material can expand consumption field of anthracite, simultaneously, offers a sustainable alternative choice for industrial fuel gas industry [7, 8, 13].

2. Feasibility of producing industrial fuel gas from anthracite

2.1 Status of industrial fuel gas in China

Fuel gas mainly includes natural gas, biogas, liquid gas, coke-oven gas, blast furnace gas, producer gas, and so on, which is widely applied in glass processing industry, ceramics production, non-ferrous metal smelting and melting, steel rolling, refractory production, etc. The source of fuel gas is multifarious and specification is complicated, but the market capacity is broad.

From the development of industrial fuel gas, many types of fuel were applied in industrial fuel gas, such as coal, electricity, heavy oil, coke oven gas, blast furnace tail gas, syngas, and natural gas. At present, the proportion of syngas (producer gas) is the largest. However, with the increasing pressure of environmental protection and publication of coal to natural gas policy, the proportion of producer gas is gradually decreasing and the proportion of natural gas is gradually increasing. **Table 1** shows specifications of fuel gases that are produced by different production technologies and with different raw materials.

The development of fuel gas derived from coal is shown as follows: first of all, UGI gasification technology, which was introduced in China in the 1930s, became the earliest technology for fuel gas derived from coal and the most widely applied technology for industrial fuel gas; because of its best adaptable to UGI gasifier, anthracite became the favorite raw material. Then, with the rise in price of anthracite, many enterprises transformed one-stage gasifier to two-stage gasifier and used low-rank lump bituminous coal as raw material to reduce production cost; simultaneously, fluidized-bed gasification technology sprung up and gradually promoted in industrial fuel gas, which led to occupy a certain proportion of the market. Now, with the development of society and improvement of environmental consciousness, Chinese government vigorously advocates the application of natural gas in fuel gas field, for its properties of being green, having low carbon, and being recognized as the cleanest fossil energy resource.

At present, problems such as low efficiency and high pollution generally exist in coal gasification technology in operation and the projects face with elimination. Many local governments require fuel gas industry to switch to natural gas, but downstream industries prefer to fuel gas derived from coal through our research. The reasons are shown as follows:

1. For its high price and unstable supply, the quantity of natural gas cannot be guaranteed at the peak of consumption.

2. For its high calorific value, high local temperature during combustion results in high content of nitrogen oxides, and the cost of denitrification is 1–3 times higher than that of coal gas.

Considering all aspects, stable supply, friendly environment, and low price are the future development directions of industrial fuel gas.

2.2 Choice of technical route by taking Lurgi gasification which used anthracite as raw material

In JAMG, the product of anthracite is primarily divided into two varieties, lump coal having particle size over 13 mm and slack coal having particle size below 13 mm [16]. Lump coal is widely used in ammonia synthesis and slightly used in civil combustion. In 2016, the sales of lump coal in coal chemistry were 1058 × 104 MT, which accounted for 80% of anthracite lump coal sales in JAMG. When new coal gasification technology has not been industrialized, anthracite lump coal was the high-quality raw material for chemical plant. For its scarcity and monopoly, anthracite lump coal had much higher price over other kinds of coal, which formed core competence for JAMG.

Since the beginning of twenty-first century, many types of gasification technology designed for other types of coal has gradually industrialized and matured, such as HT-L gasification technology, SE pulverized coal gasification technology, GSP gasification technology, shell gasification technology, opposed multi-burner coal water slurry gasification technology, and Texaco coal water slurry gasification technology. The monopolistic status of anthracite as raw material of chemical industry has been broken. Simultaneously, the gradual withdrawal of UGI furnace from chemical industry market has become a general trend and is difficult to reverse for its own technical defects. If we cannot expand new application fields, anthracite lump coal will face the market risk of falling volume and price.

Fuel gases derived from different sources		**H_2 (%)**	**CO (%)**	**CH_4 (%)**	**C2 (%)**	**C3 (%)**	**C4 (%)**	**C5+ (%)**	**Calorific value (kcal/Nm^3)**
Natural gas		—	—	97	1.5	0.5			8200–8700
Heavy oil		—	—	—	—	—	—	—	10,000–11,000 kcal/kg
Liquefied petroleum gas (LPG)	Beijing	—	—	1.5	1.0	13.5	80.2	3.8	10,800–12,000 kcal/kg
	Daqing	—	—	1.3	0.2	22.4	61.7	12.6	
Gas from oil	Catalytic cracking	58.1	10.5	16.6	5	—	—	—	4500
	Thermal cracking	31.5	2.7	28.5	26.4	5.7	—	—	10,000
Syngas from coal	Coke oven gas	59.2	8.6	23.4	—	—	—	—	4000–4500
	Converter gas	2–3	60–70	—	—	—	—	—	2000–2200
	Blast furnace tail gas	1.5–3	25–30	0.2–0.5	—	—	—	—	800–950
	One-stage gasification	7–10	23–27	1.5–3	—	—	—	—	1250–1400
	Two-stage gasification	11–15	27–31	1.5–3	—	—	—	—	1450–1600
	Oxygen enrichment gasification	30–35	30–35	1.5–2	—	—	—	—	1600–1800
	Lurgi gasification with anthracite as raw coal	35–40	25–27	7.5–9.5	—	—	—	—	2400–2600
	Fluidized-bed gasification	39	29	—	—	—	—	—	1300–2000
	Dry-pulverized gasification	19.9	66.3	—	—	—	—	—	2450–2650
	Coal water slurry gasification	35–36	44–51	—	—	—	—	—	2100–2400

Note: fuel gas derived from coal to syngas refers to crude syngas at the outlet of gasifier without any treatment.

Table 1.
Specifications of fuel gases produced by different production technologies and with different raw materials [14, 15].

The 13th Five-Year Plan of JAMG pointed out: till 2020, JAMG will build a 100 million MT coal production base and anthracite raw coal output will reach 80 million MT per year. With the gradual operation of a series of integrated coal mines, the output of anthracite lump coal will increase year by year. The mismatch between continuously increasing output of anthracite lump coal and shrinking consumption of anthracite lump coal is becoming the main contradiction, which perplexes the development of JAMG. Actively exploiting application field of anthracite lump coal has become the key to solve the above contradiction. Application as industrial fuel gas by taking Lurgi gasification, which used anthracite as raw material positively, responds to the above proposition.

The long-term, stable operation of Jinshi and Tianqing projects proved the feasibility of Lurgi gasification by using anthracite as raw material. Compared to other types of coal, Lurgi gasification by using Jincheng anthracite as raw material showed advantages including environmental protection, water consumption, effluent disposal, etc. By analyzing the characteristics of Lurgi gasification which used Jincheng anthracite as raw material, the choice of technical route was discussed and confirmed that the route of industrial gas, coal chemical products coupled with natural gas could exploit the advantage to the full.

Industrial analysis (%)				Element analysis (%)					$Q_{gr,ad}$ (MJ/kg)	$Q_{net,ad}$ (MJ/kg)	Characteristics of char residue (CRC)	Caking index (GRI)
M_{ar}	M_{ad}	A_{ad}	V_{ad}	C_{ad}	H_{ad}	O_{ad}	N_{ad}	$S_{t,ad}$				
6.12	2.46	18.98	6.54	71.78	1.90	1.92	0.52	2.44	26.92	25.53	1	0

Table 2.
Analytic index of no. 15 anthracite.

2.2.1 Operational aspect of Lurgi gasification by using anthracite as raw material

Jinshi project had a total ammonia production capacity of 30 × 10^4 MT per year, which equipped with four Lurgi gasifiers (three open, one standby). Lurgi gasifiers were Mark-IV type, which were designed by Saiding Engineering Co., Ltd. Gasifier's specification was Φ 4000 × 12,500 mm and operation pressure was 3.0 MPa. The designed dry syngas output of gasifier was 38,000 Nm^3/h, and designed raw material was no. 15 anthracite (coal quality data are shown in **Table 2**).

It was the first time in China to produce synthetic ammonium feed gas, which used Lurgi gasification technology with anthracite as designed raw material. The project started in March 2008 and implemented the whole process in May 2014. Through two-and-a-half-year operation, the feasibility of long-term and stable operation of Lurgi gasifier with anthracite as raw material was proved. When the ratio of steam to oxygen was 5.1, the components of crude syngas were shown as follows: CO 27.8%, H_2 38.4%, CH_4 7.5%, CO_2 24.8%, O2 0.2%, and others 1.3% and feed consumption was as follows: 0.53 MT raw coal/kNm^3 crude syngas, 550 kg steam/kNm^3 crude syngas, and 169 Nm^3O_2/kNm^3 crude syngas.

Based on the Jinshi project, JAMG founded Tianqing company in Qinyang city, Henan province and implemented a project that had ammonia production capacity of 30 × 104 MT and 500 million cubic meters of natural gas per year. The project equipped with six Lurgi gasifiers (five open, one standby). Lurgi gasifiers were Mark-IV type, which were designed by Saiding Engineering Co., Ltd. Gasifier's specification was Φ4000 mm × 13,000 mm, and operation pressure was 4.0 MPa. The designed dry syngas output of gasifier was 42,000 Nm^3/h, and designed raw material was no. 15 anthracite.

Since coming into operation in November 2014, the project has achieved long-term and stable operation. When the ratio of steam to oxygen was 4.1, the components of crude syngas were shown as follows: CO 23.6%, H_2 42.3%, CH_4 8.3%, CO_2 24.2%, and others 1.6%, and feed consumption was as follows: 0.45 MT raw coal/kNm^3 crude syngas, 566 kg steam/kNm^3 crude syngas, and 166 Nm^3O2/kNm^3 crude syngas.

On the basis of implementation of two above projects, JAMG set up a scientific and technical program which named "research and industrial demonstration of pressurized movable bed gasification technology for anthracite." The research and development mainly focused on gasifier structure, cooling process from chilling process to waste boiler process, and coal gas water treatment system. The program aimed to reduce project investment, reduce operation cost, improve energy efficiency, and enhance market competitiveness of Lurgi gasifier with anthracite as raw material.

2.2.2 Characteristics of Lurgi gasification by using anthracite as raw material

Compared with other types of coal, Lurgi gasification by using anthracite as raw material has the following advantages:

1. Because of high hardness, crush ratio and value loss are lower than that of other types of coal while transporting.

2. Because of low content of volatile, there are fewer refractory organic wastes such as tar and phenol in crude syngas than that of other types of coal. Some tar and phenol separation devices in traditional Lurgi gasification technology can be removed, and the investment can be reduced. There are almost no organic impurities such as tar and phenol detected in coal gas water of the outlet gasification section; the NH_3-N and COD contents in waste water are 200 and 400 mg/L, respectively [17], which induced the easiness of waste water treatment and reduction of processing cost.

3. Because of high ash melting point, the process point of gasifier is higher than that of other types of coal. Although the oxygen consumption is slightly higher than that of low-rank coal, gasification efficiency can be greatly improved. The coal gas yield of anthracite is the highest; the effective gas (which is consisted by CO, H_2, and CH_4) content ranges from 73 to 75%; and the calorific value of crude syngas is about 2500 kJ/Nm^3.

With the increase of gasification temperature, the ratio of steam to oxygen can be decreased greatly and coal gas water treatment capacity can be reduced. Simultaneously, the content of organic compounds in coal gas water will decrease and waste water will be easier to treat.

The above advantages make Lurgi gasifier with anthracite as raw material to show obvious competitiveness over that of other types of coal in environmental protection, water consumption, effluent disposal, etc.

Certainly, compared with other types of coal, Lurgi gasification by using anthracite as raw material has the following disadvantages:

1. Because of low volatile content, the content of CH_4 in crude syngas is lower than that of other types of coal.

2. Because of low volatile content, the by-products, such as tar, middle distillate, phenol, and naphtha, are lower than that of other types of coal, which lead to low economic benefits.

The above disadvantages, to a certain extent, reduce the economy of Lurgi gasification device with anthracite as raw material. So, to get advantaged position in the fierce market competition, Lurgi gasification with anthracite as raw material needs to optimize the choice of technical route and identify its own position.

2.2.3 Choice of technical route by taking Lurgi gasification with anthracite as raw material

As indicated by research findings [18], the cost of large-scale project of coal to natural gas, which takes Lurgi gasification by using anthracite as raw material, is higher than that of other types of coal, such as Datang and Qinghua projects. Under current international and domestic energy landscape, the project of anthracite coal to natural gas is basically not profitable.

Under current economic development background, technical route of Lurgi gasification with anthracite as raw material has the following choices:

1. Syngas obtained from Lurgi gasification with anthracite as raw material has advantages such as higher calorific value, better environment protection, and cheaper price than natural gas, so as to expand its application in industrial fuel gas fields, such as glass industry, ceramics industry, and nonferrous metal industry.

2. Syngas obtained from Lurgi gasification with anthracite as raw material has unique component characteristics, which should give full play and take product route of coal chemical products coupled with natural gas. Coal chemical route should take methane and ethylene glycol as products; natural gas should be sold as a high-priced non-civil product such as vehicle fuel and industrial gas in accordance with the national guidelines for the development of natural gas distributed energy.

2.3 Techno-economic evaluation of fuel gas prepared by Lurgi gasification with anthracite as raw material

2.3.1 Process flow

Through optimized design for anthracite, mainly on gasifier structure, cooling process from chilling process to waste boiler process, and coal gas water treatment system, Lurgi gasification by using anthracite as raw material can exploit the advantage and enhance market competitiveness to the full. The flow chart is shown in **Figure 1**.

2.3.2 Process index

The key factors affecting the gasification process are particle size, cohesion, mechanical strength, operating pressure, and temperature of the gasifier (**Table 3**, the process date of Jinshi and Tianqing projects).

2.3.3 Techno-economic evaluation

Based on the average price of raw material in the past 3 years (shown in **Table 4**), the economics of industrial fuel gas, which was produced by Lurgi gasification with anthracite as raw material, is discussed (shown in **Table 5**). Production capacity of crude syngas is 10×10^4 Nm^3/h. In response to this, production capacity of refined syngas is 7.4×10^4 Nm^3/h.

The result shows that total production cost of crude syngas is 653.5 ¥/kNm^3, which is equivalent to 2.56×10^{-4} ¥/kcal.

Correspondingly, total production cost of refined syngas is 960.5 ¥/kNm^3, which is equivalent to 2.78×10^{-4} ¥/kcal. Simultaneously, the average price of domestic pipeline natural gas is 2.6 ¥/Nm^3, and the equivalent cost is 3.06×10^{-4} ¥/kcal. Therefore, fuel gas produced by Lurgi gasification with anthracite as raw material shows certain advantages over natural gas in cost.

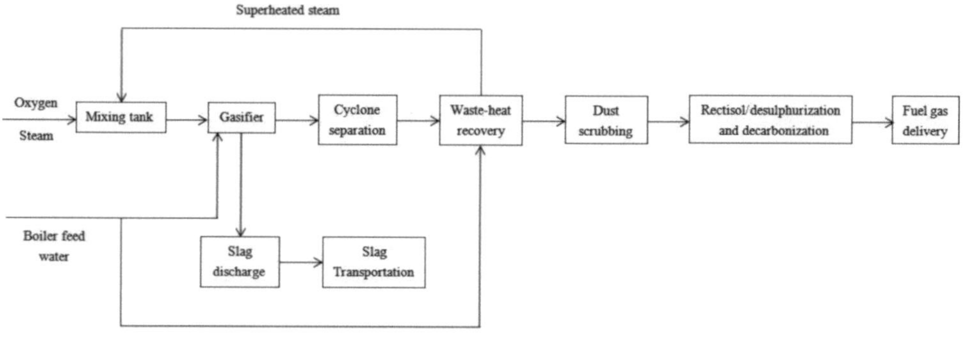

Figure 1.
Flow chart of Lurgi gasification by taking anthracite as raw material.

Name of process index		Value of process index
Type of gasifier		Mark-IV
Reactor temperature		900°C
Residence time		1 h
Process pressure		4.0 MPa
Capacity of single gasifier		500 MT/d
Particle size of raw coal		6–50 mm
Gasification agent		Oxygen and steam
Coal consumption		450 kg/kNm^3 crude syngas
Oxygen consumption		166 kNm^3/kNm^3 crude syngas
Steam consumption		402 kg/kNm^3 crude syngas
Composition of crude syngas	CO	25
	H_2	40
	CH_4	9
	CO_2	24
Calorific value of crude syngas		2550 kcal/kNm^3
Calorific value of refined syngas		3450 kcal/kNm^3

Table 3.
Process index of Lurgi gasification by taking anthracite as raw material.

Raw material	Free on board (FOB)	Freight	Average price
Anthracite	800 (¥/MT)	150(¥/MT) (vehicle distance < 300 km; train distance < 750 km)	950 (¥/MT)
Fuel coal	450 (¥/MT)	150(¥/MT) (vehicle distance < 300 km; train distance < 750 km)	600 (¥/MT)
Electricity	–	–	0.5 (¥/kwh)

Table 4.
Average price of raw material in the past 3 years.

Furthermore, fuel gas produced by Lurgi gasification with anthracite as raw material has incomparable advantages over natural gas in terms of stable supply and environment friendliness (hydrogen sulfide content is less than 1 mg/Nm^3), which fully correspond to the future development direction of industrial fuel gas industry.

2.3.4 Industrial application of fuel gas produced by Lurgi gasification with anthracite as raw material

Shahe city is the capital of flat glass in China, where production capacity of flat glass accounts for about 20% of the total production capacity in China and a lot amount of industrial fuel gas is consumed yearly. Currently, the production mode of industrial fuel gas in Shahe city basically is shown as follows: each company has its own production device of industrial fuel gas, which produces hot coal gas with calorific value of about 1500 kcal/Nm^3 by taking two-stage coal gasifier with low-rank coal as raw material. This production mode will induce serious environmental pollution for technical defects and scattered allocation. Under the background of

Index	Crude syngas as product	Refined syngas as product
Consumption of anthracite (t/kNm^3)	0.51	0.69
Cost of anthracite (¥/kNm^3)	484.5	655.5
Consumption of fuel coal (t/kNm^3)	0.12	0.2
Cost of fuel coal (¥/kNm^3)	72	120
Consumption of electricity (t/kNm^3)	50	100
Cost of electricity (¥/kNm^3)	25	50
Cost of accessories (¥/kNm^3)	5	15
Variable cost (¥/kNm^3)	586.5	840.5
Labor cost (¥/kNm^3)	10	14
Depreciation and repair costs (¥/kNm^3)	25	40
Financial cost (¥/kNm^3)	35	56
Selling expenses management cost (¥/kNm^3)	7	10
Fixed cost (¥/kNm^3)	67	120
Total cost (¥/kNm^3)	653.5	960.5

Table 5.
Economics of industrial fuel gas that was produced by Lurgi gasification with anthracite as raw material.

increasingly stringent environmental policies, a large-scale, centralized industrial fuel gas production project is planned to construct, which can supply the need of flat glass production in Shahe area and meet with the need of environment protection.

Through investigation and demonstration, Zemag Slagging Gasification (BGL) technology and Lurgi gasification technology with Shenfu coal as raw material are rejected and Lurgi gasification technology with anthracite as raw material is applied in Shahe project, which validates the technical and economic feasibility of producing industrial fuel gas by Lurgi gasification with anthracite as raw material. The scale of project is 1.6 billion Nm^3 per year, and annual consumption is 1.9 million tons of anthracite and 0.6 billion of fuel coal, respectively, which can broadly expand the utilization of anthracite. Now, the project construction is proceeding in an orderly manner.

Furthermore, our group has established extensive contacts with many projects to produce news types of fuel gas, such as glass production base located in Dezhou, Shandong province, ceramics production base located in Linyi, Shandong province, and economic development zone located in Cangzhou, Hebei province. When the conditions are ripe, our group will follow up rapidly.

3. Conclusions

Compared to natural gas, industrial fuel gas produced by Lurgi gasification with anthracite as raw material has the same environmental friendliness, while stable supply and cost are dominant, which has outstanding technical and economic feasibility. The implementation and promotion of Shahe, Dezhou, and other projects have also proved the above feasibility in practical applications.

Under current resource endowment of “rich coal, lack of oil and less gas” in China, development of industrial fuel gas produced by Lurgi gasification with anthracite as raw material has shown outstanding technical and economic feasibility, which can become an alternative to natural gas in industrial fuel gas field.

Acknowledgements

The work of this paper was accomplished as research and development engineers in the Coal Chemical Research Institute of JAMG. Thank you for financial and other support of our company, so that we can complete such a challenging research topic. In addition, other colleagues have also provided necessary assistant work for the completion of this work. Thank you.

Finally, I want to thank my beloved wife and families. With their support, I can devote myself wholeheartedly to my work and complete the writing of this research.

Conflict of interest

We declare that we do not have any commercial or associative interest that represents a conflict of interest in connection with the work submitted.

Author details

Zhang Min* and Zhang Haifeng
Shanxi Jincheng Anthracite Coal Mining Group Corporation, Ltd., Jincheng, China

*Address all correspondence to: taoismzm@gmail.com

References

[1] Congbin J, Zhu Y. Analysis on operation situation of HT-L pulverized coal pressurized gasifier and coal adaptability. Coal Processing & Comprehensive Utilization. 2014;**10**:23-28

[2] Yang Y, Xiaopeng Z, Xin G. Study on the gasification activity of raw coals for SE-gasifier pulverized coal gasification. Coal Chemical Industry. 2017;**45**(2):27-30

[3] Lixiang J. Analysis on industrial application of GSP gasification technology. Coal Engineering. 2016;**48**(1):88-91

[4] Zongfei Z, Fengjin T, Guangyou W, et al. Environmentally-friendly gasification technology-shell coal gasification process (SCGP). Chemical Fertilizer Design. 2015;**53**(6):1-4

[5] Zhenghua D, Zhijie Z, Xueli C, et al. Application of multi-opposed-burner coal water slurry gasification in chemical industry. Chemical Industry and Engineering Progress. 2006;**25**(Z1):611-615

[6] Gao L. Application of Texaco coal-water slurry pressurized gasification. Coal Technology. 2010;**29**(7):101-102

[7] diyong J, zhenhua W. Application prospect of clean industrial gas made from coal in ceramic industry. Lu Tian Hua Ke Ji. 2016;**1**:51-53

[8] zhihao H. Application prospect and development trend of coal-based industrial gas. Coal Processing&Comprehensive Utilization. 2011;**4**:51-53

[9] Haiyong S. Study on technology and economic feasibility of natural gas replaced by coal to industrial fuel gas. Coal Economic Research. 2017;**12**:25-28

[10] Zhang-wei Y, Zhi-cong W, Zong D. Present situation and development of coal chemical industry. Value Engineering. 2017;**4**:91-94

[11] Zhenqiang Z. Development of coal gasification technology. Coal Chemical Methanol. 2016;**42**(2):13

[12] Peng W, Keda Z. Advances in pressurized fixed bed gasification of crushed coal. Coal Chemical Industry. 2010;**1**:12-16

[13] Xia L, Yuanyu T, Yingyun Q. Progress of entrained-bed coal gasification technology. Chemical Industry and Engineering Progress. 2010;**29**:120-124

[14] Hanfeng M. Comparative study of different gasification technologies based on Shenfu coal. Guangzhou Chemical Industry. 2017;**45**(9):181-183

[15] Weixing Z, Yonghuang L. Analysis and comparison of typical dry pulverized coal and coal water slurry pressurized entrained-bed gasification technologies and selection. Journal of the Chemical Fertilizer Industry. 2016;**4**:40, 49-45

[16] Min Z. Discussion on choice of technical route by taking Lurgi gasification which used Jincheng anthracite as raw material. Coal Technology. 2017;**36**(10):293-294

[17] Jinzhou L, Kejia C, Chunsheng S, et al. Application of lump anthracite in crushed coal pressurized gasification process. Chemical Industry and Engineering Progress. 2016;**35**(S1):141-143

[18] Min Z. Economic assessment of coal to natural gas by Jincheng anthracite. Coal Technology. 2016;**35**(6):284-228

Chapter 6

Development of the Conical Spouted Bed Technology for Biomass and Waste Plastic Gasification

Jon Alvarez, Gartzen Lopez, María Cortazar, Laura Santamaria, Enara Fernandez and Martin Olazar

Abstract

Gasification is one of the most effective methods for upgrading different wastes, such as plastics and biomass, because the gas produced can be used directly as a fuel or as a renewable raw material for the production of chemicals and fuels. The conical spouted bed reactor (CSBR) has demonstrated to perform well in gasification process due to its specific features, such as (i) the cyclic and vigorous particle movement that avoids bed defluidization (a limitation in fluidized beds), (ii) capability for handling irregular or sticky solids, (iii) high heat transfer rates between phases, and (iv) bed stability in a wide range of gas flow rates. However, the conventional CSBR is characterized by its short residence time, which involves serious problems for minimizing tar formation. The incorporation of a fountain confiner in the CSBR is key to increasing the gas residence time and improving the contact between the gas and heat carrier particles, thereby promoting tar cracking reactions and so enhancing carbon conversion efficiency from 81.5% (without confiner) to 86.1% under fountain enhanced regime. The quality of the syngas is clearly improved as the H_2 concentration increases from 36 to 42% with and without the fountain confiner, whereas that of CO decreases from 34 to 29%, respectively.

Keywords: gasification, conical spouted bed, fountain confinement, syngas, biomass, plastic waste

1. Introduction

Gasification is a thermochemical process that transforms carbonaceous materials (coal, oil and its derivatives, biomass, post-consumer and industrial solid wastes) into syngas, with CO and H_2 being its major components. The gasification process takes place at high temperatures (generally in the 600–900°C range or even higher) in the presence of a gasifying agent (air, oxygen, steam, CO_2, or mixtures of these components) at a lower ratio than that stoichiometrically required for combustion. Syngas production is essential due to the increasing interest in gas to liquid (GTL) processes through the synthesis of methanol, dimethyl ether, and Fischer-Tropsch. In addition, the valorization of syngas can be integrated with energy recovery systems, by means of turbines, combined cycle units, or fuel cells. The gasification technology has been extensively developed for coal and oil products and is gaining increasing interest for biomass [1, 2] in which catalysts play an essential role [3]. Furthermore, the upgrading of post-consumer solid wastes by gasification is becoming a short-term promising strategy [4].

Gasification involves several steps and complex chemical reactions, which may be grouped as follows: drying, pyrolysis, cracking and reforming reactions in the gas phase, and heterogeneous char gasification. The significance of these steps on the process performance and their kinetics depends on the feedstock characteristics and gasification conditions. The pyrolysis step involves a series of complex chemical reactions of endothermic nature and leads to volatiles (gases and tars) and a solid residue or char. The homogeneous gasification reactions include a wide variety of reactions, with the balance and the extent of these reactions depending mainly on the gasifying agent used, its ratio with respect to the feed (S/feed ratio), and temperature. These reactions are as follows:

Steam reforming of hydrocarbons:

$$C_nH_m + nH_2O \rightarrow (n + m/2)\,H_2 + nCO \;\; \Delta H > 0 \qquad (1)$$

$$\text{Methane reforming:} CH_4 + H_2O \Leftrightarrow 3H_2 + CO \;\; \Delta H = 206\ \text{kJ mol}^{-1} \qquad (2)$$

$$\text{Char steam gasification:} C + H_2O \rightarrow H_2 + CO \;\; \Delta H = 131\ \text{kJ mol}^{-1} \qquad (3)$$

Dry reforming of hydrocarbons:

$$C_nH_m + nCO_2 \rightarrow (m/2)\,H_2 + 2nCO \;\; \Delta H > 0 \qquad (4)$$

$$\text{Boudouard reaction:} C + CO_2 \Leftrightarrow 2CO \;\; \Delta H = 172\ \text{kJ mol}^{-1} \qquad (5)$$

$$\text{Water - gas shift reaction:} H_2O + CO \Leftrightarrow H_2 + CO_2 \;\; \Delta H = -41\ \text{kJ mol}^{-1} \qquad (6)$$

It should be noted that gasification reactions are only those involving H_2O and CO_2, because O_2 only promotes combustion and partial oxidation reactions that produce CO, CO_2, and H_2O. In addition, the exothermic nature of oxidation reactions provides the energy required for the highly endothermic steam and CO_2 reforming (Eqs. (1)–(4)) and Boudouard (Eq. (5)) reactions. Steam improves H_2 production by means of steam reforming reactions (Eqs. (1) and (2)) and also by enhancing the water-gas shift (Eq. (6)) equilibrium. High temperatures are required for promoting char gasification, especially CO_2 gasification, whose kinetics is between 2 and 5 times slower than under steam atmosphere and does not occur below 730°C [5].

The main drawback of the syngas produced is the presence of certain impurities, such as fine particles, organic tars, NO_x, and SO_2, which need to be removed before its application in subsequent processes [6]. In particular, tar is the main contaminant in the gas produced, and its content ranges from 5 to 100 g Nm^{-3}, depending on the type of gasifier. However, its maximum allowable content is 5 mg Nm^{-3} in gas turbines and 100 mg Nm^{-3} in internal combustion engines [7, 8]. Tar is described as a complex mixture of condensable hydrocarbons, ranging from single-ring to five-ring aromatic compounds along with other oxygen-containing hydrocarbons and complex polycyclic aromatic hydrocarbons (PAHs) [9]. These compounds may cause several operational problems, such as condensation and the subsequent plugging of downstream equipment, clogging filters, and metal corrosion, which lead to unacceptable levels of maintenance for engines and turbines.

All the methods available for tar reduction may be classified into two groups, depending on where tar is removed: in situ (or primary) methods and post-gasification (or secondary) methods. Regardless of the strategy followed, the optimum operating conditions, appropriate additives or catalysts, and a suitable reactor configuration should be established in order to obtain a gas stream with a maximum tar content of 2 g Nm^{-3} and a low content of PAH compounds [10, 11]. It should be noted that tar formation depends on the gasification conditions, particularly on temperature, so preventive treatments are recommended to operate above 1000°C. Aznar et al. [12] suggest injecting a secondary air stream into the freeboard to reduce the content of tar.

In situ catalytic cracking is one of the most promising techniques, as it allows reducing the need for expensive downstream operations [3, 13]. Natural minerals, such as olivine [14, 15] and dolomite [16], have been widely used in steam gasification because, apart from being active for the cracking and reforming of heavy aromatic compounds, they are inexpensive and abundant. In addition, Ni catalysts have received great attention in gasification due to their higher effectiveness for converting tar into H_2-rich gas [17, 18].

Moreover, apart from temperature and catalysts, reactor design also plays a critical role in gasification. Different reactor configurations are commonly used for the steam gasification process, which according to their hydrodynamic behavior can be classified as follows: fixed bed, fluidized bed, entrained flow, and rotary kiln reactors, among others [4]. Fluidized beds are the most commonly used due to their advantages, such as versatility for using different types of wastes (agroforestry, post-consumer, and industrial), high heat and mass transfer rates between phases, and bed isothermicity, which allow the scaling-up of the process to the industrial level [19–21]. Nevertheless, biomass or waste particles of irregular texture require a large amount of inert solid (sand) to promote their fluidization. In addition, small particle sizes (Geldart A and B) are the best for fluidization, and therefore high amounts of energy are required to grind and sieve the feedstock. Nevertheless, there is an alternative to conventional fluidized beds, namely, the conical spouted bed reactor (CSBR), which may handle residues of different densities and sizes without significant segregation in the bed. This technology allows handling larger particles than those in fluidized beds, including those with an irregular texture, fine materials, and sticky solids, with no agglomeration or segregation problems [22]. Moreover, the highly vigorous movements of the solids lead to high heat and mass transfer rates between phases [23]. Other advantages of the CSBR over the fluidized bed are its simpler design (no distributor plate) and the lower sand/feed ratio required for the same capacity.

The main drawback of this technology for gasification is the short gas residence time, which hinders tar cracking reactions. Accordingly, certain modifications have been developed in order improve its performance in the gasification process by changing reactor hydrodynamics, which are as follows: the confinement of the fountain and the use of draft tubes. The fountain confinement device is a tube welded to the lid of the reactor that allows operating under stable conditions with fine particles and increasing the gas residence time by lengthening the path followed by the gas [24]. Therefore, gas-solid (catalyst) contact in the fountain is greatly improved, and tar cracking and reforming reactions are therefore promoted. Moreover, the draft tube also enables to widen the operation range and improve the reactor's hydrodynamic behavior [24]. Thus, this chapter summarizes the main results obtained in the application of the conical spouted bed reactor in the steam gasification of biomass and waste plastics. Moreover, the influence of different primary catalysts and the incorporation of novel modifications in the reactor design, such as fountain confiner and draft tube, are also discussed.

2. Material and methods

2.1 Feedstock characterization

The HDPE was supplied by Dow Chemical (Tarragona, Spain) in the form of chippings (4 mm), with the following properties: average molecular weight, 46.2 kg mol^{-1}; polydispersity, 2.89; and density, 940 kg m^{-3}.

The biomass used in this study is forest pinewood waste (*Pinus insignis*). The sawdust has been sieved to obtain a particle size between 1 and 2 mm. This material has been dried at room temperature to a moisture content below 10 wt%. Ultimate and proximate analyses have been carried out in a LECO CHNS-932 elemental analyzer and in a TGA Q500IR thermogravimetric analyzer, respectively. The high heating value (HHV) for both biomass and HDPE was measured in a Parr 1356 isoperibolic bomb calorimeter. The main features of both the raw biomass and the HDPE are summarized in **Table 1**.

2.2 Equipment

Steam gasification runs have been carried out in a bench-scale plant, whose scheme is shown in **Figure 1**. The main element of the plant is the conical spouted bed reactor (CSBR), whose design is based on previous hydrodynamic studies [25] and on the application of this technology to the pyrolysis of different solid wastes, such as biomass [26–28], plastics [29], and waste tires [30].

	Biomass	HDPE
Ultimate analysis (wt%)		
Carbon	49.33	85.71
Hydrogen	6.06	14.29
Nitrogen	0.04	—
Oxygen	44.57	—
Proximate analysis (wt%)		
Volatile matter	73.4	99.7
Fixed carbon	16.7	0.3
Ash	0.5	—
Moisture	9.4	—
HHV (MJ kg^{-1})	19.8	43.1

Table 1.
Characterization of the biomass and HDPE used in this study.

The plant is provided with a system for continuous feeding of the biomass or plastic. The system for solid feeding consists of a vessel equipped with a vertical shaft connected to a piston placed below the material bed. The plastic/sawdust is fed into the reactor by raising the piston at the same time as the whole system is vibrated by an electric engine.

Water has been fed by means of a Gibson 307 pump that allows a precise measuring of the flow rate. The water stream has been vaporized by means of an electric cartridge placed inside the forced convection oven and prior to the entrance of the reactor.

The reactor is located within an oven, which is in turn placed in a forced convection oven maintained at 270°C to avoid the condensation of steam and tars before the condensation system. A high-efficiency cyclone and a sintered steel filter (5 μm) are also placed inside this oven in order to retain the fine sand particles entrained from the bed and the soot or char particles formed in the gasification process.

The gases leaving the forced convection oven circulate through a volatile condensation system consisting of a condenser, a Peltier cooler, and a coalescence filter. The Peltier cooler consists of a 150 mL tank and a refrigerator that lowers the temperature to around 2°C, thereby efficiently condensing the volatile products. The condenser is a double-shell tube cooled by tap water.

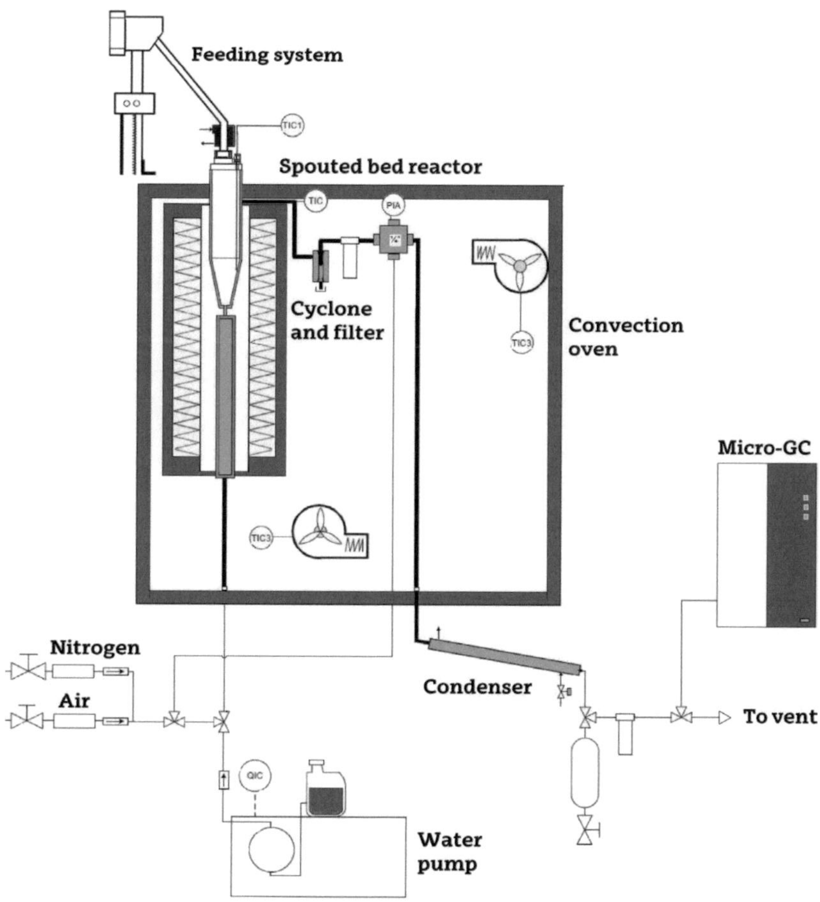

Figure 1.
Scheme of the bench-scale biomass gasification plant equipped with a conical spouted bed reactor.

2.2.1 Conventional spouted bed reactor

The spouted bed reactor is the core of the gasification plant. The total height of the reactor is 298 mm, with that of the conical section (angle of 30°) being 73 mm. The diameters of the cone base and cylindrical section are 12.5 and 60.3 mm, respectively. The gas inlet diameter is 7.6 mm. Despite the endothermic nature of the steam gasification process, bed isothermicity is ensured by the vigorous solid circulation of the sand in this reactor, which also promotes high heat transfer rates [23]. The CSBR is placed inside a 1250 W radiant oven. Two K-type thermocouples are located inside the reactor, one in the bed annulus and the other one close to the wall.

2.2.2 Fountain-enhanced spouted bed reactor

This reactor is an improved version of that described in Section 2.2.1, which has been specifically designed for gasification process. Thus, a fountain confiner was welded to the lid in order to increase the residence time, narrow its distribution, and improve the gas-solid contact in the fountain region (**Figure 2**). Thus, several

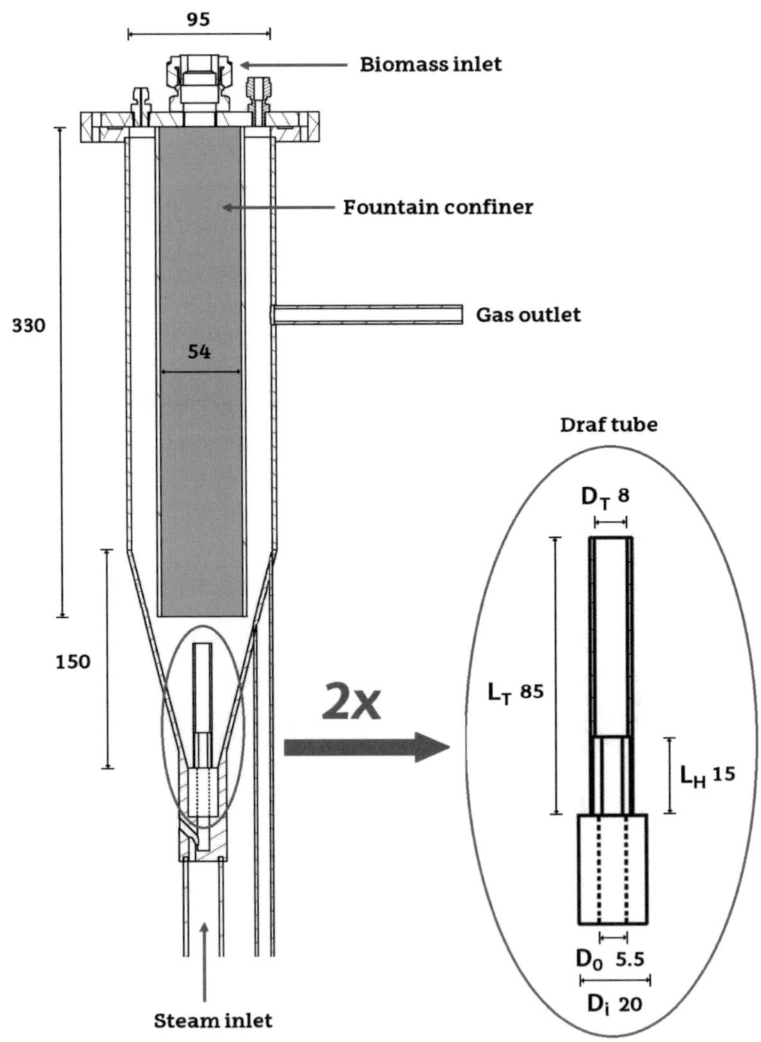

Figure 2.
Main dimensions (in mm) of the spouted bed gasifier, fountain confiner, and draft tube.

modifications were introduced in order to optimize its performance. For example, the height of the reactor was increased in order to increase the residence time of the gas and promote tar cracking. This reactor may also operate in the conventional spouting regime by using a lid without confiner. It is noteworthy that its design allows using draft tubes to widen the application range of the spouting regime and improve bed stability [31, 32]. In fact, the nonporous draft tube promotes high fountains [32] by diverting most of the inlet gas stream through the draft tube, which also enhances solid cross-flow from the annulus into the spout and therefore leads to additional gas-solid contact in the fountain.

The main dimensions of this spouted bed reactor, the fountain confiner, and the draft tube used are depicted in **Figure 2**. According to a previous hydrodynamic study conducted under gasification conditions [33], a draft tube with 8 mm in external diameter (5.5 mm in internal diameter) and 15 mm entrainment zone height was determined as the optimum one. Thus, these geometric factors allow operating under enhanced fountain regime, with low steam flow rates ensuring great turbulence and a well-developed fountain region with a great hydrodynamic stability.

2.3 Primary catalysts

γ-Al_2O_3 has been provided by Alfa Aesar and olivine by Minelco. Olivine has been calcined at 900°C for 10 h prior to use in the gasification reaction to enhance its reactivity for tar cracking. The conditions mentioned for olivine calcination have been determined as optimum by Devi et al. [14] in order to maximize tar cracking activity. The BET surface area has been measured by N_2 adsorption-desorption (Micromeritics ASAP 2010). Calcined olivine has a limited porosity, with a surface area of only 0.18 $m^2\ g^{-1}$. However, γ-Al_2O_3 has a much higher porous development, with a surface area of 159 $m^2\ g^{-1}$.

2.4 Product analysis

The volatile stream leaving the gasification reactor has been analyzed online by means of a GC Agilent 6890 provided with a HP-PONA column and a flame ionization detector (FID). The sample has been injected into the GC by means of a line thermostated at 280°C, once the reactor outlet stream has been diluted with an inert gas. The purpose of this system is to avoid the condensation of tars in the transfer line. The tars collected in the condensation system have been identified in a gas chromatograph/mass spectrometer (GC/MS, Shimadzu UP-2010S provided with a HP-PONA column). The non-condensable gases have been injected into a micro-GC (Varian 4900).

2.5 Experimental procedure

Temperature and steam/biomass ratio are the operating parameters studied in the gasification of biomass and plastics in this reactor. Additionally, biomass gasification was also performed with different primary catalysts (in situ), and the influence of using the fountain confiner was evaluated. In all runs, water flow rate was 1.5 mL min^{-1}, corresponding to a steam flow rate of 1.86 L min^{-1}, which is approximately 1.5 times that corresponding to the minimum spouting velocity in order to ensure bed stability.

The effect of temperature has been studied at 800, 850, and 900°C by feeding a mass flow rate of 1.5 g min^{-1} of biomass or HDPE and using a steam/feed ratio of 1.

The effect of the steam/feed ratio has been studied between 0 and 2 (in mass), and the temperature has been maintained at 900°C. For a ratio of 2, the biomass or plastic feed rate was reduced to 0.75 g min^{-1} in order to maintain the same steam flow rate (1.5 g min^{-1}). The reactor contains 70 g of sand in the bed in all runs, and therefore the residence time of the products in the reactor and the hydrodynamic behavior are similar. In order to study the steam/feed ratio of 0, the steam was replaced with a N_2 flow rate of 2 L min^{-1}.

In the experiments to assess the effect of the primary catalyst on product distribution, the bed contains 70 g of sand or olivine (with particle diameter in the 0.35–0.4 mm range). However, given that γ-Al_2O_3 has a much lower density, the bed of this material contained 25 g with a particle size greater than that of sand, in the 0.4–0.8 mm range, in order to attain a similar hydrodynamic behavior in all cases. The experiments were carried out at 900°C, with a feed rate of 1.5 g min^{-1} of HDPE or sawdust and with a steam/feed ratio of 1.

In the experiments performed with the fountain-confined spouted bed, the biomass feed rate was 0.75 g min^{-1}, with a steam/biomass ratio of 2. The bed contained 100 g of olivine, and two particles sizes have been used, i.e., 90–150 and 250–355 μm. These olivine particle size ranges are those corresponding to the optimum hydrodynamic performance of the reactor, as the minimum spouting velocity depends strongly on particle size [33]. Thus, the gas velocity in the runs with the coarse olivine fraction corresponds to approximately 1.5 times the minimum spouting velocity (so the reactor operated under conventional spouting regime), whereas in the experiments performed with the fine olivine, the gas velocity used is approximately four times higher than the minimum spouting velocity (4 u_{ms}), and the fountain-enhanced regime was therefore attained.

Furthermore, operation was carried out in two regimes in the same reactor in order to ascertain the influence the confinement system (in the standard spouting regime) has on the biomass gasification process. Thus, experiments with and without the fountain confiner were carried out at 850°C and S/B of 2, using coarse olivine (250–355 μm), with gas velocity corresponding in both cases to approximately 1.5 times u_{ms} (conventional spouting regime). The results obtained with the confiner under conventional spouting regime were compared with those obtained with this device but operating in the enhance fountain spouting regime under the same conditions and replacing the coarse olivine with the fine one in the bed. Therefore, the role of the vigorous gas-catalyst contact in the fountain-enhanced regime was assessed.

All the runs were performed in continuous mode for 20 min in order to ensure a steady-state process. The char yield was determined by weighing the mass in the reactor, as well as those retained in the cyclone and in the sintered steel filter. The char yield is given by mass unit of the whole amount of solid fed into the reactor (approximately 30 g). All the runs have been repeated several times (at least three) under the same conditions in order to guarantee reproducible results.

3. Results and discussion

3.1 HDPE gasification

In this work, steam gasification of HDPE has been studied in the conventional conical spouted bed pilot plant described in Section 2.2. The effect of temperature (in the 800–900°C range) and steam/plastic (S/P) ratio (between 0 and 2) on the gas yield, tar content, carbon conversion efficiency, and H_2 production is shown in **Table 2**. The reaction indices have been defined as follows: (i) gas yield as the volumetric gas production (on a dry basis) per kg of biomass in the feed (on a wet basis), (ii) tar yield expressed as the tar mass per syngas m^3 (on a dry basis), (iii) carbon conversion efficiency as the ratio between the carbon units contained in the syngas and those contained in the biomass in the feed, and (iv) H_2 production as the mass percentage of the H_2 produced per biomass mass unit.

The gaseous fraction is composed of H_2, CO, and CO_2, together with C_2–C_5 hydrocarbons (mainly C_3–). The tar is defined as the amount of organic compounds with a molecular weight and boiling point higher than that of benzene, a criterion that is commonly used by most authors [11, 14, 34]. The char is a carbonaceous product collected after the reaction in the reactor, sintered steel filter, and cyclone. The mass balance closure in all the experiments was above 95%.

As observed in **Table 2**, an increase in temperature leads to higher gas yields and lower tar and char yields, thus improving the efficiency of the whole process. The gas yield increases from 2.5 $m^3\ kg^{-1}$ of HDPE at 800°C to 3.4 $m^3\ kg^{-1}$ of HDPE at 900°C. Furthermore, the carbon conversion efficiency at 800°C is 86%, increases to 91% at 850°C, and then remains constant with further increases in temperature to 900°C.

Tar content decreases from 29.5 g Nm^{-3} at 800°C to 16.7 g Nm^{-3} at 900°C due to the enhancement of thermal cracking. Other authors have also observed a positive effect of temperature on the tar cracking in the gasification of waste plastics by using both steam [35] and air [36] as gasifying agents. In fact, according to certain authors, the destruction of tar aromatic hydrocarbons only occurs at temperatures above 850°C [13].

The influence of temperature on product yields has also been studied with different gasification technologies, and most of the authors agree that higher temperatures enhance syngas yield and decrease that of tar and char [37–39]. Higher char yields than those shown in **Table 2** have been reported in the literature [35, 40], which may be attributed to the characteristics of the gas-solid contact in the conical spouted bed reactor, which mitigate the limitations in the physical steps prior to gasification, which are as follows: (i) plastic melting, (ii) coating of sand particles, and (iii) pyrolysis.

Table 2 also displays the reaction indices for different S/P values. As observed, as S/P ratio is increased from 1 to 2, the carbon conversion increases from 91.0 to 93.6%. Note that the performance is poor when operating with a S/P = 0 (pyrolysis), given that carbon conversion efficiency is as low as 68.6% due to the high tar and char yields. The lack of steam in the reactor at high temperatures promotes the formation of aromatic compounds, leading to a tar content as high as 29.5 g Nm^{-3}. The presence of steam in the reaction medium increases the gas yield and decreases that of tar. When operating only with N_2 as a fluidizing agent, the tar concentration

Temperature (°C)	S/P ratio	Tar content (g Nm^{-3})	Carbon conversion (%)	Gas yield ($m^3\ kg^{-1}$)	H_2 production (wt%)	Char yield (wt%)
800	1	29.5	86.1	2.5	12.7	1.4
850	1	13.8	91.1	3.2	17.0	0.6
900	1	16.7	91.1	3.4	18.4	0.5
900	2	9.6	93.6	3.6	19.9	0.4
900	0	207.8	68.6	0.9	2.7	5.6

Table 2.
Effect of gasification temperature and S/P ratio on the gas yield, tar content, carbon conversion efficiency, and H_2 production.

is 207.8 g Nm^{-3}, but this concentration is drastically reduced to 16.7 g $Nm^{-3}$and 9.6 g Nm^{-3} when operating with S/P ratios of 1 and 2, respectively. These results suggest that an increase in S/P ratio enhances the cracking of tar compounds, as reported by Herguido et al. [41] in the steam gasification of biomass.

The presence of steam in the reaction environment also improves H_2 production, increasing significantly from 2.7 to 18.4 wt% when the S/P ratio is increased from 0 to 1. However, the increase in H_2 production (19.9 wt%) is moderate when a S/P value of 2 is used. Similarly, gas yield increases slightly from 3.4 m^3 kg^{-1} HDPE to 3.6 m^3 kg^{-1} HDPE when the S/P ratio is raised from 1 to 2. The following aspects can explain these results: (i) promotion of hydrocarbon reforming reactions (Eq. (1)) as steam concentration is higher and (ii) low tar and char formation rate, although this effect is of lower significance. A similar trend has been reported in the literature, although some authors attain a saturating trend, i.e., a higher steam/tire ratio than the optimum one does not increase the gas yield [42, 43].

Moreover, **Figure 3** displays the composition of the gases formed at different temperatures (**Figure 3a**) and S/P ratios (**Figure 3b**). As observed in **Figure 3a**, an increase in temperature leads to an increase in the concentrations of H_2, CO, and CH_4 in the gaseous stream, which are 60.3, 28.2, and 7.2% vol., respectively, at 900°C. Temperature has an opposite effect on C_2–C_5 hydrocarbons (made up mainly of olefins, with ethylene being the major one), whereas that on CO_2 was almost negligible (the concentration is almost steady).

The higher concentration of H_2 and CO can be explained by the endothermic nature of steam and dry reforming reactions (Eqs. (1) and (4)), which are promoted at higher temperatures, whereas that of CH_4 is due to the endothermicity of HDPE cracking reactions. On the contrary, the C_2–C_5 hydrocarbons formed are probably reformed, and therefore their yield decreases as temperature is higher. It should be noted that the water-gas shift reaction (Eq. (6)) is exothermic, and therefore thermodynamic equilibrium shifts toward the formation of CO at high temperatures.

Regarding the gas composition (**Figure 3b**), an increase in S/P ratio from 1 to 2 does not lead to a significant change, but the composition of the gas when only pyrolysis is performed (S/P = 0) is very different. As observed, the presence of steam favors H_2 and CO_2 formation but reduces that of CO and CH_4 because the higher concentration of steam in the reactor enhances both water-gas shift and methane reforming reactions. Other authors have observed a similar effect of S/P ratio on the gas composition in the gasification of different polymeric materials [42, 43].

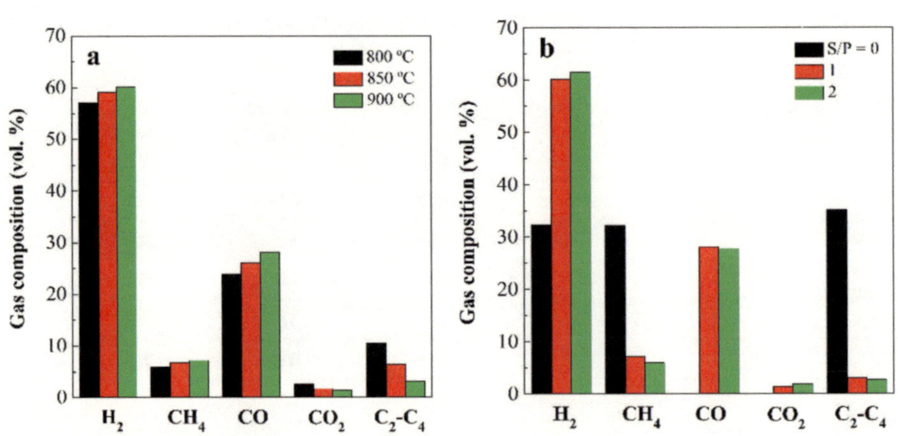

Figure 3.
Effect of gasification temperature (a) and S/P ratio (b) on the gaseous fraction composition.

3.2 Biomass gasification

3.2.1 Effect of temperature and S/B ratio

The same reaction indices in plastic gasification, i.e., the gas yield, tar content, carbon conversion efficiency, and H_2 production, have been analyzed in this section (**Table 3**). Temperature is one of the more influential variables in steam gasification, and its effect has been studied in the 800–900°C range for a steam/biomass ratio of 1. Moreover, the effect of steam/biomass ratio has been studied in the 0–2 range at 900°C. Note that no steam was fed into the reactor in the runs carried out with a S/B ratio of 0, but the sawdust contained a moisture content of approximately 10%, and water is formed during the thermal degradation of biomass. Therefore, some steam reforming will occur even in the runs without water in the feed (runs with S/B = 0).

As observed in **Table 3**, the temperature plays a crucial role in the efficiency of the gasification process. An increase in the gasification temperature reduces the tar content in the gaseous product from 364.4 g Nm^{-3} at 800°C to 142.5 g Nm^{-3} at 900°C. The gas yield also increases from 0.7 m^3 kg^{-1} of biomass at 800°C to 1 m^3 kg^{-1} of biomass at 900°C, whereas that of char decreases from 8.9% at 800°C to 4.5% at 900°C. In the same line, the carbon conversion efficiency in the process is considerably higher as temperature is increased, and char yield is therefore lower. In fact, char gasification kinetics is enhanced by temperature due to the highly endothermic nature of char steam gasification (Eq. (3)) and Boudouard (Eq. (4)) reactions. The increase in char conversion with temperature is related to the shift in equilibrium in both reactions [44]. However, this result depends on the char residence time in the reactor. Thus, char gasification reaction kinetics is slow, even above 800°C.

Although the content of tar is reduced to 142.5 g Nm^{-3} operating at 900°C due to the positive effect of temperature on tar cracking and reforming reactions, this value is still high for syngas applications. It should be noted that no defluidization problems are observed in the steam gasification, which is due to the vigorous solid cyclic movement in the conical spouted bed. However, the conventional spouted bed regime leads to short residence times (below 0.5 s), which are beneficial to increase the yield of bio-oil in pyrolysis processes, but in gasification they are responsible for the limited tar cracking, whose concentration in the gaseous stream is rather high, as observed in **Table 3**.

Given that the tar yield is highly dependent on several parameters, such as residence time, temperature, and S/B ratio, the results showed in the literature vary greatly depending on the technology used, but all of them evidence a significant decrease in tar content in the gaseous product stream with temperature [45–47].

With respect to the experiments carried out with different S/B ratios (**Table 3**), an increase in this parameter improves the gasification performance by increasing the gas yield and carbon efficiency and lowering that of tar. For example, tar concentration has been reduced from 154 g Nm^{-3} with a S/B = 0 to 142.5 g Nm^{-3} with a S/B = 1, given that an increase in the S/B ratio promotes tar cracking and reforming reactions (Eq. (1)). However, a further increase in the S/B ratio from 1 to 2 only reduces slightly the tar content of the gaseous product. Likewise, the gas yield increases from S/B 0 to 1 (from 0.9 to 1 m^3 kg^{-1} of biomass) but hardly changes as S/B is increased from 1 to 2.

The reduction in the tar and char content leads to an increase in the carbon conversion efficiency, attaining the maximum value of 70% with a S/B = 2. Although gasification efficiency is improved in terms of biomass conversion, the energy efficiency of the process is lower when high S/B ratios are used, given that more water

Temperature (°C)	S/P ratio	Tar content (g Nm^{-3})	Carbon conversion (%)	Gas yield (m^3 kg^{-1})	H_2 production (wt%)	Char yield (wt%)
800	1	364.2	50.4	0.7	1.9	8.9
850	1	243.1	59.1	0.8	2.5	6.3
900	1	142.5	69.8	1.0	3.2	4.5
900	2	142.0	70.0	1.0	3.6	3.6
900	0	154.0	50.4	0.9	2.3	10.7

Table 3.
Effect of gasification temperature and S/B ratio on product fraction yields, carbon conversion efficiency, and tar concentration, at 900°C.

need to be vaporized and the unreacted steam needs to be recovered after being condensed. Kaushal and Tyagi [48] suggest optimum S/B ratios between 0.6 and 0.85, which guarantee the thermal efficiency of the process and, at the same time, the presence of enough steam in the gasifier to promote steam reforming reactions.

The composition of the gases (on a dry basis) formed at different temperatures and different S/B ratios is displayed in **Figure 4**. As observed in **Figure 4a**, an increase in temperature enhances H_2 formation due to the endothermic nature of the reactions involved (Eqs. (1)–(5)). Moreover, the inorganic species of the biomass retained in the char have a positive effect on the water-gas shift reaction (Eq. (6)) at higher temperatures [46]. Accordingly, H_2 concentration increases from 28% at 800°C to 38% at 900°C, whereas that of CO decreases from 41.5 to 32.5% in the same range of temperature. Besides, concentration of methane and the other gaseous hydrocarbons (C_2 to C_4) decreases as temperature is raised due to the enhancement of hydrocarbon reforming reactions. As in HDPE gasification, the effect of temperature on CO_2 is not of significance, as its concentration increases slightly between 800 and 900°C.

Figure 4b shows the composition of the gaseous stream for different S/B ratios. Given that the WGS reaction and methane and hydrocarbon reforming reactions (Eqs. (1) and (2)) are promoted at high S/B ratios, the formation of H_2 and CO_2 is enhanced, whereas that of CO and hydrocarbons is hindered. It is to note that this effect is more remarkable when the S/B ratio is increased from 0 to 1.

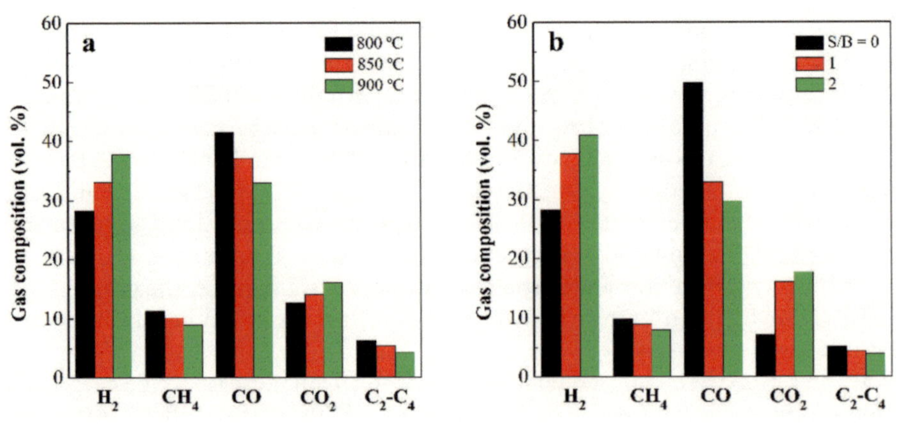

Figure 4.
Gas composition (on a dry basis) for the steam gasification at different temperatures (a) and S/B ratios (b).

3.2.2 Effect of primary catalyst

The experiments with different bed materials have been carried out at a temperature of 900°C and a S/B ratio of 1. **Table 4** shows the effect of the primary catalysts used (olivine and γ-alumina) on reaction indices (gas yield, tar content, H_2 production, and carbon conversion) and compares the results with those obtained using inert sand as bed material. As observed, both olivine and γ-alumina cause a great decrease in tar content (30.1 and 22.4 g Nm^{-3}, respectively) compared to the runs carried out with inert sand (142.5 g Nm^{-3}). Accordingly, both catalysts improve the gasification performance, with tar reduction being slightly higher for γ-alumina (84%) than that for olivine (79%). Moreover, the carbon conversion efficiency has a drastic increase when a primary catalyst is used, attaining a value of 86.8% for olivine and 87.6% for γ-alumina. It is noteworthy that H_2 production peaks at 4.5 wt% when the γ-alumina is used.

As mentioned above, tar formation leads to operational problems in the gasification and subsequent units for syngas processing; thus, the use of a catalyst, such as olivine and γ-alumina, improves process efficiency, especially the latter, which significantly reduces tar content. Nevertheless, olivine is cheaper and more available because it is a natural material [49]. Other papers in the literature also report considerable improvements in gasification efficiency by using primary catalysts [45, 50].

The effect primary catalysts have on gas composition is displayed in **Figure 5**. As observed, γ-alumina has a greater influence on gas composition than olivine. The presence of catalysts leads to an increase in H_2 and CO_2 concentrations and a

	Sand	Olivine	γ-Alumina
Tar content (g Nm^{-3})	142.5	30.2	22.4
Carbon conversion (%)	69.7	86.8	87.6
Gas yield (m^3 kg^{-1})	1.0	1.1	1.2
H_2 production (wt%)	3.2	3.7	4.5
Char yield (g Nm^{-3})	4.5	4.3	4.3

Table 4.
Effect of the primary catalysts on reaction indices.

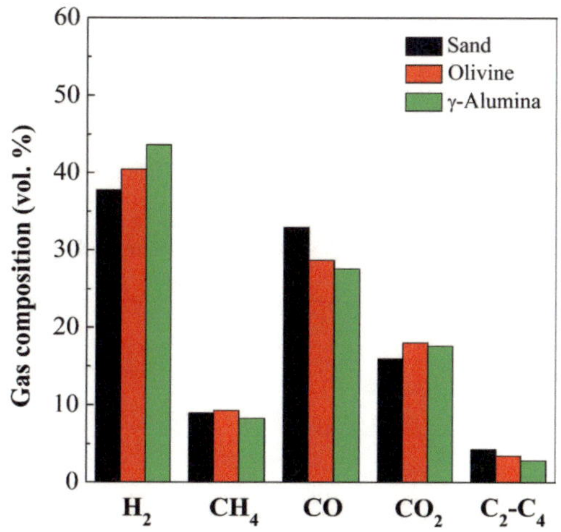

Figure 5.
Effect of primary catalysts on the composition of the gaseous fraction.

reduction in that of CO due to the promotion of the water-gas shift reaction (Eq. (6)). In addition, the higher concentration of H_2 by the presence of this type of catalyst is also related to the enhancement of tar cracking and reforming reactions (Eq. (1)). Moreover, γ-alumina also seems to promote methane and light hydrocarbon reforming (Eq. (3)), which can be deduced from their lower concentration in the presence of this catalyst.

3.2.3 Effect of fountain confinement on biomass gasification

Runs have been carried out with a S/B ratio of 2 and at a temperature of 850°C with different spouting regimes and gas flow patterns developed in conical spouted beds, such as (i) standard spouting regime without fountain confiner, (ii) standard spouting regime with fountain confiner, and (iii) enhanced fountain regime with fountain confiner. **Table 5** compares the gas yield, tar content, carbon conversion efficiency, char yield, and H_2 production results obtained for the three configurations mentioned.

As observed in **Table 5**, the incorporation of the fountain confiner leads to a decrease in tar content in the syngas from 49.2 g Nm^{-3} without fountain confiner to 34.6 g Nm^{-3} when this device is inserted. The volatiles in the conventional spouted bed gasifier leave quickly from the reaction zone through the outlet located in the gasifier upper section. Thus, the short residence time of the volatiles limits the contact of tars and other gaseous products with the catalyst, which hinders cracking and reforming reactions and therefore lowers conversion efficiency. On the contrary, the fountain confiner prevents the premature leaving of the gases at an initial stage in the biomass gasification and causes a downward gas flow inside the confiner, which favors the contact between the volatile stream and the catalyst. Furthermore, the confined fountain and the use of draft tubes lead to a highly stable hydrodynamic regime, which allows operating with finer materials (lower particle sizes of olivine) and higher fountain heights [24].

In order to analyze the influence on the gasification performance by changing the gas-catalyst contact in the reactor, especially in the fountain region, runs with the fountain confiner were performed under similar residence times (same reactor geometry and gas flow rate) as in conventional conical spouted beds. As observed in **Table 5**, the promotion of steam reforming of tars and gaseous hydrocarbons using the confinement system improved the gas yield and H_2 production from 1.1 to 1.2 m^3 kg^{-1} and from 3.5 to 4.6 wt%, respectively. In the same line, the carbon conversion efficiency also increased when the confinement system was used, given that a value of 83.6% was obtained instead of 81.5% without this system. It should be remarked that these values are slightly higher than those reported by other authors in fluidized bed reactors under similar conditions [51, 52].

	Without confiner	With confiner (standard spouting)	With confiner (enhanced fountain)
Tar content (g Nm^{-3})	49.2	34.6	20.6
Carbon conversion (%)	81.5	83.6	86.1
Gas yield (m^3 kg^{-1})	1.1	1.2	1.3
H_2 production (wt%)	3.5	4.6	5.0
Char yield (g Nm^{-3})	6.5	6.2	6.0

Table 5.
Influence of the confinement system and spouting regime on the reaction indices.

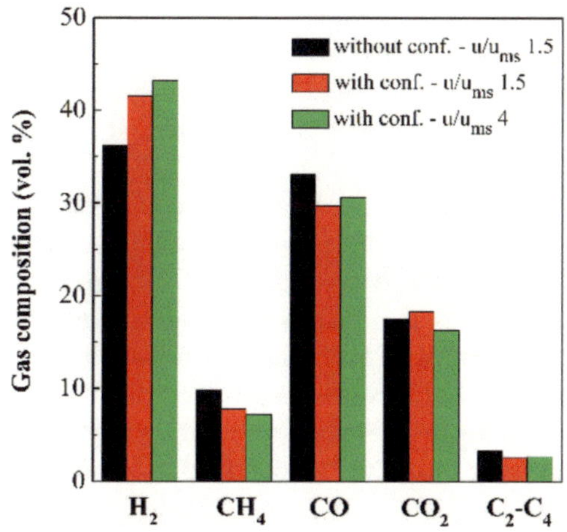

Figure 6.
Influence of the confinement system and spouting regime on gas composition.

Table 5 also shows that the results are greatly improved under fountain-enhanced regime by decreasing olivine particle size and increasing the fountain height. In fact, the tar content in the gas is reduced from 34.6 g Nm^{-3} under conventional spouting regime up to 20.6 g Nm^{-3} under enhanced fountain regime. This improvement is associated with the better gas-catalyst contact and heat transfer rates in the fountain region due to the higher fountain height. Furthermore, the smaller particle size of olivine increased the catalyst surface area available for cracking and reforming reactions [14]. Moreover, gas composition with and without confiner (under conventional and fountain-enhanced regime) is shown in **Figure 6**.

As observed in **Figure 6**, H_2 concentration increases from 36 to 42% with and without the fountain confiner, whereas that of CO decreases. The effect on CO_2 is not so remarkable, but its concentration is slightly higher when the fountain confiner is introduced. Furthermore, the concentration of methane and the other gaseous hydrocarbons decreased due to the higher extent of steam reforming reactions involving methane (Eq. (2)) and tar (Eq. (1)), as well as of water-gas shift (Eq. (6)) reactions when the fountain confiner was used. This improvement is related to the increase in the gas residence time and the better contact of the gas with the catalyst attained when the fountain confiner is used. It is noteworthy that effect of the fountain-enhanced regime on the gas composition is rather limited. The most significant change is that regarding H_2 concentration, whose value increases to 43.2%.

4. Conclusions

The conical spouted bed reactor is an interesting technology for the continuous steam gasification of biomass and waste plastics due to the high heat transfer rates for a highly endothermic process (as is gasification) as well as to the absence of defluidization problems. An increase in gasification temperature improves process efficiency in terms of conversion to gases, with the maximum carbon conversion being of 70 and 91.1% at 900°C for biomass and HDPE, respectively. Furthermore, steam/feed ratio has a positive effect on the composition of the gas by increasing the H_2 concentration from 32 to 61% in the HDPE gasification and from 28 to 42% in that of biomass when steam/feed ratio is increased from 0 to 2. In fact, higher steam concentrations in the reaction environment enhance both tar cracking and char gasification and so increase carbon conversion efficiency. Nevertheless, the concentration of the tars attained is still high for its direct application. The use of primary catalysts, such as olivine and γ-alumina, has shown an excellent performance for tar elimination as their content is being reduced by up to 30.1 and 22.4 g Nm^{-3} with olivine and γ-alumina, respectively.

The incorporation of a fountain confiner in the CSBR allows modifying bed hydrodynamics, i.e., increase the residence time of the volatiles and improve their contact with the catalyst in order to promote gasification performance and favor tar cracking. Hence, H_2 productions and carbon conversion efficiencies increase when the fountain confiner is introduced from 3.5 to 4.6 wt% and from 81.5 to 83.6%, respectively. Moreover, the H_2 concentration increases from 36 to 42%, whereas that of CO decreases from 34 to 29% with and without the fountain confiner. This device allows operating under enhanced fountain regime by reducing olivine particle size, which leads to a better contact between olivine and the gases, and therefore tar content is further reduced, and the carbon conversion efficiency increases up to 86.1%.

Acknowledgements

This work was carried out with financial support from the Ministry of Economy and Competitiveness of the Spanish Government (CTQ2016-75535-R (AEI/FEDER, UE) and CTQ-2015-69436-R (MINECO/FEDER, UE)), the European Regional Development Funds (ERDF), the Basque Government (IT748-13), and the University of the Basque Country (UFI 11/39).

Author details

Jon Alvarez[1], Gartzen Lopez[2*], María Cortazar[2], Laura Santamaria[2], Enara Fernandez[2] and Martin Olazar[2]

1 Department of Chemical and Environmental Engineering, University of the Basque Country UPV/EHU, Vitoria-Gasteiz, Spain

2 Department of Chemical Engineering, University of the Basque Country UPV/EHU, Bilbao, Spain

*Address all correspondence to: gartzen.lopez@ehu.es

References

[1] Damartzis T, Zabaniotou A. Thermochemical conversion of biomass to second generation biofuels through integrated process design—A review. Renewable and Sustainable Energy Reviews. 2011;**15**:366-378

[2] Segurado R, Pereira S, Correia D, Costa M. Techno-economic analysis of a trigeneration system based on biomass gasification. Renewable and Sustainable Energy Reviews. 2019;**103**:501-514

[3] Shahbaz M, Yusup S, Inayat A, Patrick DO, Ammar M. The influence of catalysts in biomass steam gasification and catalytic potential of coal bottom ash in biomass steam gasification: A review. Renewable and Sustainable Energy Reviews. 2017;**73**:468-476

[4] Arena U. Process and technological aspects of municipal solid waste gasification. A review. Waste Management. 2012;**32**:625-639

[5] Di Blasi C. Combustion and gasification rates of lignocellulosic chars. Progress in Energy and Combustion Science. 2009;**35**:121-140

[6] Font Palma C. Modelling of tar formation and evolution for biomass gasification: A review. Applied Energy. 2013;**111**:129-141

[7] Zhou B, Dichiara A, Zhang Y, Zhang Q, Zhou J. Tar formation and evolution during biomass gasification: An experimental and theoretical study. Fuel. 2018;**234**:944-953

[8] Anis S, Zainal ZA. Tar reduction in biomass producer gas via mechanical, catalytic and thermal methods: A review. Renewable and Sustainable Energy Reviews. 2011;**15**:2355-2377

[9] Cortazar M, Alvarez J, Lopez G, Amutio M, Santamaria L, Bilbao J, et al. Role of temperature on gasification performance and tar composition in a fountain enhanced conical spouted bed. Energy Conversion and Management. 2018;**171**:1589-1597

[10] Rios MLV, Gonzalez AM, Lora EES, Almazan O. Reduction of tar generated during biomass gasification: A review. Biomass and Bioenergy. 2018;**108**:345-370

[11] Karl J, Pröll T. Steam gasification of biomass in dual fluidized bed gasifiers: A review. Renewable and Sustainable Energy Reviews. 2018;**98**:64-78

[12] Aznar MP, Caballero MA, Sancho JA, Francés E. Plastic waste elimination by co-gasification with coal and biomass in fluidized bed with air in pilot plant. Fuel Processing Technology. 2006;**87**:409-420

[13] Devi L, Ptasinski KJ, Janssen FJJG. A review of the primary measures for tar elimination in biomass gasification processes. Biomass and Bioenergy. 2003;**24**:125-140

[14] Devi L, Craje M, Thune P, Ptasinski KJ, Janssen F. Olivine as tar removal catalyst for biomass gasifiers: Catalyst characterization. Applied Catalysis, A: General. 2005;**294**:68-79

[15] Tian Y, Zhou X, Lin S, Ji X, Bai J, Xu M. Syngas production from air-steam gasification of biomass with natural catalysts. Science of the Total Environment. 2018;**645**:518-523

[16] Gusta E, Dalai AK, Uddin MA, Sasaoka E. Catalytic decomposition of biomass tars with dolomites. Energy and Fuels. 2009;**23**:2264-2272

[17] Artetxe M, Alvarez J, Nahil MA, Olazar M, Williams PT. Steam reforming of different biomass tar model compounds over Ni/Al_2O_3

catalysts. Energy Conversion and Management. 2017;**136**:119-126

[18] Caballero MA, Aznar MP, Gil J, Martín JA, Francés E, Corella J. Commercial steam reforming catalysts to improve biomass gasification with steam-oxygen mixtures. 1. Hot gas upgrading by the catalytic reactor. Industrial and Engineering Chemistry Research. 1997;**36**:5227-5239

[19] Xiong Q, Yeganeh MM, Yaghoubi E, Asadi A, Doranehgard MH, Hong K. Parametric investigation on biomass gasification in a fluidized bed gasifier and conceptual design of gasifier. Chemical Engineering and Processing. 2018;**127**:271-291

[20] Molino A, Chianese S, Musmarra D. Biomass gasification technology: The state of the art overview. Journal of Energy Chemistry. 2016;**25**:10-25

[21] Sansaniwal SK, Pal K, Rosen MA, Tyagi SK. Recent advances in the development of biomass gasification technology: A comprehensive review. Renewable and Sustainable Energy Reviews. 2017;**72**:363-384

[22] Aguado R, Alvarez S, San José MJ, Olazar M, Bilbao J. Gas flow distribution modelling in conical spouted beds. Computer Aided Chemical Engineering. 2005;**20**:613-618

[23] Makibar J, Fernandez-Akarregi AR, Alava I, Cueva F, Lopez G, Olazar M. Investigations on heat transfer and hydrodynamics under pyrolysis conditions of a pilot-plant draft tube conical spouted bed reactor. Chemical Engineering and Processing. 2011;**50**:790-798

[24] Altzibar H, Estiati I, Lopez G, Saldarriaga JF, Aguado R, Bilbao J, et al. Fountain confined conical spouted beds. Powder Technology. 2017;**312**:334-346

[25] Olazar M, San José MJ, Aguayo AT, Arandes JM, Bilbao J. Design factors of conical spouted beds and jet spouted beds. Industrial and Engineering Chemistry Research. 1993;**32**:1245-1250

[26] Alvarez J, Hooshdaran B, Cortazar M, Amutio M, Lopez G, Freire FB, et al. Valorization of citrus wastes by fast pyrolysis in a conical spouted bed reactor. Fuel. 2018;**224**:111-120

[27] Alvarez J, Lopez G, Amutio M, Bilbao J, Olazar M. Bio-oil production from rice husk fast pyrolysis in a conical spouted bed reactor. Fuel. 2014;**128**:162-169

[28] Alvarez J, Amutio M, Lopez G, Barbarias I, Bilbao J, Olazar M. Sewage sludge valorization by flash pyrolysis in a conical spouted bed reactor. Chemical Engineering Journal. 2015;**273**:173-183

[29] Artetxe M, Lopez G, Elordi G, Amutio M, Bilbao J, Olazar M. Production of light olefins from polyethylene in a two-step process: Pyrolysis in a conical spouted bed and downstream high-temperature thermal cracking. Industrial and Engineering Chemistry Research. 2012;**51**:13915-13923

[30] Alvarez J, Lopez G, Amutio M, Mkhize NM, Danon B, van der Gryp P, et al. Evaluation of the properties of tyre pyrolysis oils obtained in a conical spouted bed reactor. Energy. 2017;**128**:463-474

[31] Altzibar H, Lopez G, Bilbao J, Olazar M. Minimum spouting velocity of conical spouted beds equipped with draft tubes of different configuration. Industrial and Engineering Chemistry Research. 2013;**52**:2995-3006

[32] Makibar J, Fernandez-Akarregi AR, Díaz L, Lopez G, Olazar M. Pilot scale conical spouted bed pyrolysis reactor: Draft tube selection and hydrodynamic

performance. Powder Technology. 2012;**219**:49-58

[33] Lopez G, Cortazar M, Alvarez J, Amutio M, Bilbao J, Olazar M. Assessment of a conical spouted with an enhanced fountain bed for biomass gasification. Fuel. 2017;**203**:825-831

[34] Schmid JC, Wolfesberger U, Koppatz S, Pfeifer C, Hofbauer H. Variation of feedstock in a dual fluidized bed steam gasifier-influence on product gas, tar content, and composition. Environmental Progress & Sustainable Energy. 2012;**31**:205-215

[35] Wu C, Williams PT. Pyrolysis-gasification of plastics, mixed plastics and real-world plastic waste with and without Ni-Mg-Al catalyst. Fuel. 2010;**89**:3022-3032

[36] Arena U, Zaccariello L, Mastellone ML. Fluidized bed gasification of waste-derived fuels. Waste Management. 2010;**30**:1212-1219

[37] Niu Y, Han F, Chen Y, Lyu Y, Wang L. Experimental study on steam gasification of pine particles for hydrogen-rich gas. Journal of the Energy Institute. 2017;**90**:715-724

[38] Michel R, Rapagna S, Di Marcello M, Burg P, Matt M, Courson C, et al. Catalytic steam gasification of *Miscanthus X giganteus* in fluidised bed reactor on olivine based catalysts. Fuel Processing Technology. 2011;**92**:1169-1177

[39] Barisano D, Canneto G, Nanna F, Alvino E, Pinto G, Villone A, et al. Steam/oxygen biomass gasification at pilot scale in an internally circulating bubbling fluidized bed reactor. Fuel Processing Technology. 2016;**141**:74-81

[40] He M, Xiao B, Hu Z, Liu S, Guo X, Luo S. Syngas production from catalytic gasification of waste polyethylene: Influence of temperature on gas yield and composition. International Journal of Hydrogen Energy. 2009;**34**:1342-1348

[41] Herguido J, Corella J, González-Saiz J. Steam gasification of lignocellulosic residues in a fluidized bed at a small pilot scale. Effect of the type of feedstock. Industrial and Engineering Chemistry Research. 1992;**31**:1274-1282

[42] Donatelli A, Iovane P, Molino A. High energy syngas production by waste tyres steam gasification in a rotary kiln pilot plant. Experimental and numerical investigations. Fuel. 2010;**89**:2721-2728

[43] Wu C, Williams PT. Hydrogen production from the pyrolysis-gasification of polypropylene: Influence of steam flow rate, carrier gas flow rate and gasification temperature. Energy and Fuels. 2009;**23**:5055-5061

[44] Baratieri M, Baggio P, Fiori L, Grigiante M. Biomass as an energy source: Thermodynamic constraints on the performance of the conversion process. Bioresource Technology. 2008;**99**:7063-7073

[45] Rapagna S, Jand N, Kiennemann A, Foscolo PU. Steam-gasification of biomass in a fluidised-bed of olivine particles. Biomass and Bioenergy. 2000;**19**:187-197

[46] Wei L, Xu S, Zhang L, Liu C, Zhu H, Liu S. Steam gasification of biomass for hydrogen-rich gas in a free-fall reactor. International Journal of Hydrogen Energy. 2007;**32**:24-31

[47] Umeki K, Yamamoto K, Namioka T, Yoshikawa K. High temperature steam-only gasification of woody biomass. Applied Energy. 2010;**87**:791-798

[48] Kaushal P, Tyagi R. Steam assisted biomass gasification-an overview.

Canadian Journal of Chemical Engineering. 2012;**90**:1043-1058

[49] Erkiaga A, Lopez G, Amutio M, Bilbao J, Olazar M. Steam gasification of biomass in a conical spouted bed reactor with olivine and γ-alumina as primary catalysts. Fuel Processing Technology. 2013;**116**:292-299

[50] Koppatz S, Pfeifer C, Hofbauer H. Comparison of the performance behaviour of silica sand and olivine in a dual fluidised bed reactor system for steam gasification of biomass at pilot plant scale. Chemical Engineering Journal. 2011;**175**:468-483

[51] Franco C, Pinto F, Gulyurtlu I, Cabrita I. The study of reactions influencing the biomass steam gasification process. Fuel. 2003;**82**:835-842

[52] Carpenter DL, Bain RL, Davis RE, Dutta A, Feik CJ, Gaston KR, et al. Pilot-scale gasification of corn stover, switchgrass, wheat straw, and wood: 1. Parametric study and comparison with literature. Industrial and Engineering Chemistry Research. 2010;**49**:1859-1871